Studia Fennica
Historica 16

The Finnish Literature Society was founded in 1831 and has from the very beginning engaged in publishing. It nowadays publishes literature in the fields of ethnology and folkloristics, linguistics, literary research and cultural history.
The first volume of Studia Fennica series appeared in 1933.
Since 1992 the series has been divided into three thematic subseries: Ethnologica, Folkloristica and Linguistica. Two additional subseries were formed in 2002, Historica and Litteraria. Subseries Anthropologica was formed in 2007.
In addition to its publishing activities the Finnish Literature Society maintains a folklore archive, a literature archive and a library.

Editorial board
Markku Haakana
Pekka Hakamies
Timo Kaartinen
Pauli Kettunen
Leena Kirstinä
Hanna Snellman
Kati Lampela

Editorial office
Hallituskatu 1
FIN-00170 Helsinki

Sport, Recreation and Green Space in the European City

Edited by Peter Clark, Marjaana Niemi & Jari Niemelä

Finnish Literature Society • Helsinki

Studia Fennica Historica 16

The publication has undergone a peer review.

The open access publication of this volume has received part funding via a Jane and Aatos Erkko Foundation grant.

© 2009 Peter Clark, Marjaana Niemi, Jari Niemelä and SKS
License CC-BY-NC-ND 4.0. International

A digital edition of a printed book first published in 2009 by the Finnish Literature Society.
Cover Design: Timo Numminen
EPUB Conversion: Tero Salmén

ISBN 978-952-222-162-9 (Print)
ISBN 978-952-222-791-1 (PDF)
ISBN 978-952-222-790-4 (EPUB)

ISSN 0085-6835 (Studia Fennica)
ISSN 1458-526X (Studia Fennica Historica)

DOI: http://dx.doi.org/10.21435/sfh.16

This work is licensed under a Creative Commons CC-BY-NC-ND 4.0. International License.
To view a copy of the license, please visit http://creativecommons.org/licenses/by-nc-nd/4.0/

A free open access version of the book is available at http://dx.doi.org/10.21435/sfh.16 or by scanning this QR code with your mobile device.

Contents

PREFACE... 7

Peter Clark, Marjaana Niemi and Jari Niemelä
1   INTRODUCTION 9

PART I: NATIONAL TRENDS IN SPORTS AREAS DEVELOPMENT

Katri Lento
2   A QUESTION OF GENDER, CLASS AND POLITICS:
    THE USE AND PROVISION OF SPORTS GROUNDS
    IN HELSINKI C. 1880S–1960S............................ 25

Suvi Talja
3   SPORT FOR ALL? – DEVELOPMENT OF SPORTS SITES
    AND GREEN SPACE IN HELSINKI SINCE THE 1960S 41

Pim Kooij
4   URBAN GREEN SPACE AND SPORT: THE CASE
    OF THE NETHERLANDS, 1800–2000 58

Christiane Eisenberg
5   PLAYING FIELDS IN GERMAN CITIES, 1900–2000 76

Fulvia Grandizio
6   GREEN SPACE AND SPORT IN ITALIAN CITIES IN THE
    TWENTIETH CENTURY: THE EXAMPLE OF TURIN *90*

PART II: SPORTS AREAS, PLANNING AND ENVIRONMENT

Jürgen H. Breuste
7 GREEN SPACE, PLANNING AND ECOLOGY IN GERMAN CITIES
    IN THE LATE TWENTIETH CENTURY 113

Jarmo Saarikivi
8 ECOLOGICAL PERSPECTIVES OF SPORTS GREEN SPACE
    IN THE CONTEMPORARY EUROPEAN CITY 125

Suvi Talja
9 GOLF AND GREEN SPACE IN FINLAND: TALI GOLF COURSE
    SINCE 1932 143

Philip James and Emma L. Gardner
10 THE ROLE OF PRIVATELY OWNED SPORTS RELATED GREEN
     SPACES IN URBAN ECOLOGICAL FRAMEWORKS 161

Christiane Eisenberg and Reet Tamme
11 THE GOLF BOOM IN GERMANY 1980–2006:
     COMMERCIALISATION, NATURE PROTECTION
     AND SOCIAL EXCLUSION 171

Niko Lipsanen
12 CHANGING PLACES OF SPORT IN THE LIGHT
     OF PHOTOGRAPHS: HELSINKI 1976 & 2006 184

EPILOGUE

Jussi S. Jauhiainen
13 GREEN SPACE AND SPORT: GOLF, PARKOUR AND OTHER
     POST-DISCIPLINARY OPPORTUNITIES 199

LIST OF FIGURES AND TABLES 209
NOTES ON CONTRIBUTORS 212
INDEX 214

Preface

This volume is one of the major outcomes of an innovative interdisciplinary project on *Green Space, Sport and the City* based at the University of Helsinki between 2005–2007 and led by Professor Peter Clark (Department of History) and Professor Jari Niemelä (Department of Environmental Sciences) in collaboration with Dr Marjaana Niemi (Department of History and Philosophy, University of Tampere). The project was funded in 2005–2006 by the University of Helsinki European Studies Network with grants for two researchers for two years, and support for project meetings. We are grateful to the Network for its support. Thanks are also due to the Nessling Foundation for a grant to finance an international workshop in 2006 and to contribute to the costs of this volume. The City of Helsinki kindly made a grant towards the cost of some of the research for the project. We are also grateful to the Finnish Literature Society and the Editorial Board of the Studia Fennica series for including this book in their series.

Special thanks are due to Professor Pauli Kettunen (as the editor of the Studia Fennica Historica series) and two anonymous referees for their valuable comments and suggestions. Finally, we would like to thank Rauno Endén, Mikko Rouhiainen and Kati Lampela at the Finnish Literature Society for seeing the book through the publishing process.

Helsinki 20 November 2009

Peter Clark
Jari Niemelä
Marjaana Niemi

PETER CLARK, MARJAANA NIEMI AND JARI NIEMELÄ

1 Introduction

Sports areas comprise a significant part of the European urban landscape – for example, in London around 14 per cent of the urban space – and in some cities they are more extensive than nature reserves. They are potentially a major ecological resource and contribute significantly to urban sustainability. Since the end of the nineteenth century the growth of sports sites has had major consequences not only for the urban environment but for urban planning, municipal policy and public health. While the European sport and sports ground revolution was launched in Britain in the second half of the nineteenth century, by the First World War British-type sports had spread to much of continental Europe and the Nordic countries. Yet, as we shall see in this book, the pattern of sports ground development during the twentieth century was highly varied across Europe, with major differences between countries. Regionality is clearly of major significance in examining environmental changes and impacts, as we show here.[1]

In this volume 12 contributors from Finland, Britain, the Netherlands, Germany, Austria and Italy examine the contested development of urban green spaces and sports sites from a comparative and transdisciplinary viewpoint, bringing together the expertise of historians, geographers and ecologists. Such a transdisciplinary approach reflects the growing concern of environmental studies with the idea of political ecology – the extent to which environmental changes and social processes are intertwined and interact. As Alf Hornborg and others have recently argued: 'In integrating cultural, political, economic, and ecological perspectives… political ecology requires transdisciplinary analyses that handle the great variety of factors involved…'[2]

Since its beginnings, political ecology has primarily been concerned with rural development questions such as erosion, deforestation and desertification and with questions of ecological processes on a large scale, mainly in

---

1 See also A. Hornborg, 'Environmental history as political ecology', in A. Hornborg, J.R. McNeill and J. Martinez-Alier (eds), *Rethinking Environmental History: World-System History and Global Environmental Change.* Lanham: Altamira Press 2007, 1.
2 Hornborg, 'Environmental history', 3.

developing countries.³ In recent years, however, interest in urban environment has been growing, and political ecology has addressed many questions that are also central to urban history. One of the clearest connections between urban political ecology and urban history is an interest in tracing linkages between emergent urban forms and previous ones, including built and 'natural' elements. Both disciplines examine changing historical assumptions and ideas about the role of nature in the urban context – as a source of aesthetic pleasure, instrument of moral and social reform, place of encounter and sociability, potential site for redevelopment, site for ecological restoration and rehabilitation – and the ways these assumptions and ideas are linked to power relations and socio-economic processes such as urbanisation, industrialisation and de-industrialisation. As Stephanie Pincetl argues, the ways in which urbanisation or industrialisation 'have transformed the natural environment and appropriated and reallocated space and nature are indicative of relations of power as well as ideas of nature'.⁴

The chapters in this book shed light on several key themes central to urban political ecology and urban history: the evolution, creation and use of sports and recreation areas and their relationship to the changing assumptions about the role of nature in urban form; the way in which green or other sports areas have shaped and been shaped by gender and class relations; and the way in which the creation and use of these areas have reflected and reinforced changes in sports and leisure activities. The chapters also address the role of different actors, especially city governments but also state agencies and various associations, in the planning and management of sports areas. Finally, the book offers new insights into the environmental significance of sports areas in the wider debate about urban green spaces. One type of sports site which figures in a number of the following chapters is the golf course, which in recent decades has seen dynamic growth in many European countries.

Recreational areas and sports grounds represented, as we can see from Figure 1.1, only one new category of urban green space in the modern European city – along with the many different types of parks, allotment gardens, cemeteries, company gardens and the like. But the relationship of sport and green space has been a particularly controversial one, as many of the case studies below illustrate and as Jussi Jauhiainen reiterates in the wide-ranging concluding chapter. There are questions about the impact of organised sport including golf on urban ecology. Has it been destructive of urban biodiversity or have the effects been more variable? What is the impact

---

3   S. Pincetl, 'The political ecology of green spaces in the city and linkages to the countryside', *Local Environment* 12 (2007) 2, 87–92. See also P. Robbins, *Political Ecology: A Critical Introduction*. Malden MA: Blackwell Publishing 2004, 5–12.
4   Pincetl, 'The political ecology of green spaces', 89. See also J.P. Evans, 'Wildlife corridors: an urban political ecology' and J. Byrne, M. Kendrick and D. Sroaf, 'The park made of oil: towards a historical political ecology of the Kenneth Hahn State Recreation Area', *Local Environment* 12 (2007) 2, 129–152 and 153–181; M. Gandy, *Concrete and Clay: Reworking Nature in New York City*. Cambridge MA and London: MIT Press 2003. For urban history, see for example P. Clark (ed.), *The European City and Green Space: London, Stockholm, Helsinki and St Petersburg 1850–2000*. Aldershot: Ashgate 2006.

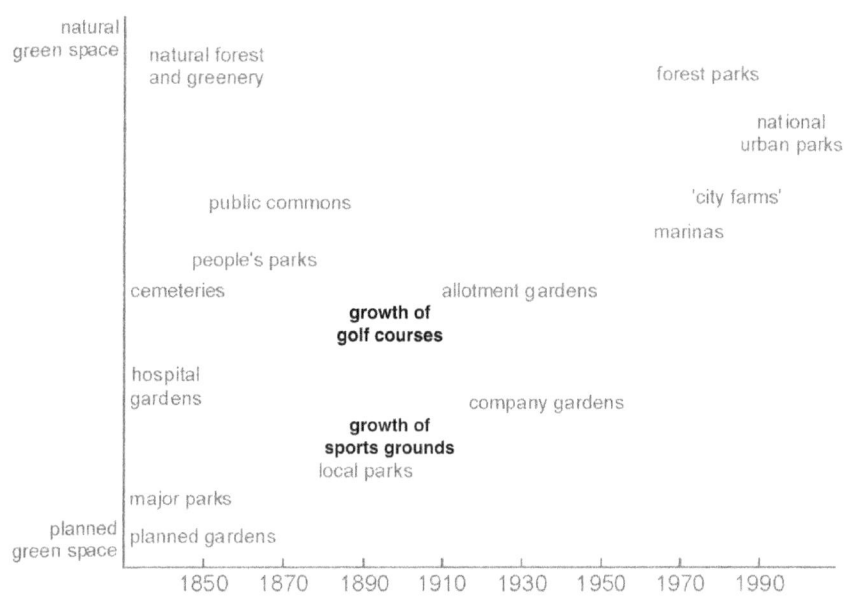

Figure 1.1 The development of sports areas and green spaces in European Cities

of sport on the construction and use of green space – has it promoted the contesting and segregation of spaces, disadvantaging non-sportsmen, women, and ethnic minorities, or has it a more social integrative role? Again what is the relationship of sport and public space – has it tended to encourage the privatisation of public space? The needs and impacts of different sports also need to be explored, whether golf, football or parkour.

In our introduction we try and spell out the broad framework of the development of sports spaces in European cities from the end of the nineteenth century to the present time, outlining in turn sports trends, the role of urban policy, and issues of environmental impact, before summarising briefly the case study chapters that follow.

Overall trends in sports spaces

So far the development of sports areas in modern European cities has attracted only limited research, and there is a need for more multidisciplinary study. Whereas sports activities, particularly organised competitive and commercial sports, have received widespread attention – for instance in Britain by John Lowerson, Wray Vamplew, Tony Mason, and others,[5] in Finland by Henrik

---

5   J. Lowerson, *Sport and the English Middle-Classes 1870–1914*. Manchester: Manchester University Press, 1993; W. Vamplew, *Pay up and Play the Game: Professional Sport in Britain 1875–1914*. Cambridge: Cambridge University Press 1988; T. Mason, *Association Football and English Society 1863–1915*. Brighton: Harvester Press 1980, on soccer.

Meinander,[6] and in Germany by Christiane Eisenberg, Hajo Bernett and Michael Krüger,[7] those extensive spaces in which such games and activities were performed have been frequently ignored. Despite the extensive literature on municipal parks in the nineteenth century, investigating garden styles, park design and social and educational functions, relatively little has been written about their evolution as sports and recreation venues.[8] John Allan Patmore's work in 1970 studied the recreational use of urban parks in the 1960s in the context of national changes in leisure and land-use and some recent work has pointed to the growth of sports sites in the general development of urban green space.[9] Research has also started to appear on individual sporting venues and on the growth of commercial stadia.[10] However, only a handful of scholars have examined the broad range of sports sites in cities. The Danish social scientist Henning Eichberg has argued for an alternation of indoor and outdoor arenas for sport since the nineteenth century, while the geographer John Bale has proposed for Britain for a more steady growth of sports spaces linked to national identity and the growing commercialism and internationalism in sport.[11]

All sports have their own distinctive profile and social scientists have identified over a hundred recognised sports, but as the following chapters suggest sports can be categorised in four broad types: organised amateur sports, essentially team sports; international competitive sports; commercial sports; and informal, individualistic sports. Linked to these different categories, it is arguable that the growth of sports areas in European cities has followed three main phases, though with major national variations.

---

6 H. Meinander, *Towards a Bourgeois Manhood: Boys Physical Education in Nordic Secondary Schools 1880–1940*. Helsinki: Finnish Society of Science and Letters 1994; S. Aalto, 'Urheilu – viihteestä viihdeteollisuudeksi', in *Helsingin historia vuodesta 1945. 2.* Helsinki: City of Helsinki 2000, 341–485.

7 C. Eisenberg, *"English sports" und deutsche Bürger: Eine Gesellschaftsgeschichte 1800–1939.* Paderborn: Ferdinand Schöningh 1999; H. Bernett, *Leichtathletik im geschichtlichen Wandel.* Schorndorf: Hoffmann 1987; M. Krüger, *Körperkultur und Nationsbildung: Die Geschichte des Turnens in der Reichsgründungsära – eine Detailstudie über die Deutschen.* Schorndorf: Hoffmann 1996.

8 H. Conway, *People's Parks: The Design and Development of Victorian Parks in Britain.* Cambridge: Cambridge University Press 1991, devotes only a few pages to sports grounds; though Catharina Nolin's study of Stockholm parks *Till Stadsbornas nytta och förlustande.* Stockholm, Byggförlaget 1999, has more on the new changes. M. Hannikainen, Park life: an urban environmental history of Battersea Park, 1846–1951 (Unpublished Masters thesis, History Department, University of Helsinki, 2005) is an excellent study of the transformation of municipal parks for sport.

9 J.A. Patmore, *Land and Leisure.* London: David and Charles 1970.

10 For attempts to position the development of sports sites in the general development of urban green spaces see several of the chapters in Clark (ed.), *The European City*; see also P. Clark, S. Jokela and J. Saarikivi, 'Nature, sport and the European city: London and Helsinki 1880–2005' (paper at a conference on 'Nature in the City', German Historical Institute, Washington, December 2005), to be published in S. Dümpelmann and D. Brantz (eds), *The Place of Nature in the City* (forthcoming).

11 J. Bale and C. Philo (eds), *Henning Eichberg. Body Cultures: Essays on Sport, Space and Identity.* London: Routledge 1998; J. Bale, *Landscapes of Modern Sport.* Leicester: Leicester University Press 1994.

The first phase, from the later nineteenth century up to the 1920s, was a time when many organised sports were established and spread, mainly on an amateur basis. As noted, Britain led the way. Though some organised elite sports had existed earlier, the mid Victorian era was a watershed. Traditional football was transformed, largely through the medium of the public schools, into two organised games – football (soccer) and rugby: the *Football Association* was formed in 1863 and the *Rugby Football Union* in 1871. Hockey spread quickly in the 1870s as a middle-class game with the *Hockey Association* established in 1886. Athletics developed from the 1860s with the *Amateur Athletic Club* founded by 1866. Golf advanced strongly in late Victorian Britain as did lawn tennis.[12] A similar if later chronology of the establishment of sports is evident in the Netherlands (see Chapter Four) and other European countries.[13] Across Europe, the growth of organised sports reflected strong demand created by urbanisation, rising incomes, increased leisure time and a greater awareness of sport as vital for health. There was also important elite and bourgeois support, as the ruling classes saw organised sport as a means of promoting new notions of masculinity, national identity, military education, and the improvement of the lower classes, linked to ideas of eugenics.

With the expansion of organised amateur sports came a slow growth of outdoor sports and recreation sites, with pitches marked out in more or less standardised geometric form, often situated in public parks. By the 1890s London County Council controlled over 400 cricket pitches, 300 tennis courts, and a hundred or so football pitches, but generally the provision of public sports facilities was not a high priority. The erection of stadia for commercial and international sports likewise made only limited progress before the First World War.[14]

The second main phase of development from the 1920s up to the 1960s saw a massive proliferation of sports venues for organised amateur activity, including recreation grounds in old and new municipal parks, specialist sports grounds, school and factory sports fields, outdoor swimming pools, as well as indoor facilities. Demand was strong as a result of the mass popularisation of sports after the First World War linked to better living standards, further reductions in working hours, a new attitude to nature and improved public (and later private) transport. In the post-war period demand for sports facilities accelerated, and there was growing if still lagging participation by women. In the late 1960s over a third of male Londoners (and 16 per cent of women) said they played sport regularly or occasionally, and in the next decade about half of the residents of Helsinki were actively engaged in outdoor sports. The period also saw a significant expansion of large stadia for commercial and international sports (such as the Olympic Games) which

---

12 H. Cunningham, *Leisure in the Industrial Revolution c.1780 – c.1880.* London: Croom Helm 1980, 114 et seq.; Lowerson, *Sport and the English Middle-Classes*, 73, 79, 81.
13 See below Chapters by Pim Kooij, Christiane Eisenberg and Fulvia Grandizio.
14 Hannikainen, Park life: an urban environmental history of Battersea Park; Clark, Jokela and Saarikivi, 'Nature, sport and the European city'.

attracted growing media attention and so helped stimulate public interest in sport generally.[15]

The last main phase of development in sports sites occurred from the 1970s and 1980s with the rapid expansion of informal, individualistic and commercial sports as well as the growing competition from rural pursuits, and, partly as a consequence, the relative decline of amateur organised sport. There was increased provision for informal sports (climbing walls, skating parks, jogging tracks, orienteering areas) alongside the expansion of commercial venues (football stadia, health clubs). Most striking has been the escalating number of golf courses. There was a major take-off in Britain in the 1960s and from the 1980s courses multiplied in other European countries, usually run by private companies and in some cases by international chains. By comparison, facilities for amateur and international sports may have generally stagnated reflecting reduced public use and interest and changes in municipal policy and planning.[16]

This phased chronology of development tells only part of the story, however. As several of the chapters below reveal, there were important national and even local variations in the growth and type of provision of sports grounds and other facilities. Whilst the Netherlands generally followed Britain during the twentieth century in the growth of mainly outdoor, 'green' sports grounds, in Germany the trend was more towards 'grey' sports facilities, including artificial tracks and indoor sports complexes, at least until the last decades of the twentieth century. In Finland there was a combination of grey and green sites, and also 'white' facilities because of the long snowy winters. In Italy the growth of public sports grounds in the twentieth century was apparently meagre, with most facilities owned by the elite or private companies.[17]

As we noted earlier, even where it exists, sports space – like all urban space – may be heavily contested. Sports areas may exclude many more than they admit: local residents kept out from private golf courses, lower-class people unable to afford membership of private sports clubs or pay the access charges for municipal facilities; women discriminated against by heavily segregated, male-dominated sports, at least before the last decade of the twentieth century.[18] Sports activists have often set the agenda for the design of recreational space. Niko Lipsanen points out in Chapter Twelve how even municipal sports facilities in Helsinki appeared increasingly fenced, turning into segregated spaces reserved for sportsmen. If demand, including

---

15 Clark, Jokela and Saarikivi, 'Nature, sport and the European city'; K. Lento, 'The role of nature in the city: green space in Helsinki, 1917–60', in Clark (ed.), *The European City*, 189 et passim; P.L. Garside, 'Politics, ideology and the issue of open space in London 1939–2000', in Clark (ed.), *The European City*, 71; *Surveys of the Use of Open Spaces: 2 vols*. London: Greater London Council, 1968–72, 1.
16 Patmore, *Land and Leisure*; Clark, Jokela and Saarikivi, 'Nature, sport and the European city'; Bale, *Landscapes of Modern Sport*, 49 et seq., 84.
17 See below Chapters Two – Six.
18 For a recent polemic on gender inequality in sport: E. McDonagh and L. Pappano, *Playing with the Boys: Why Separate is not Equal in Sports*. Oxford: Oxford University Press 2008. This has little on the spatial dimension.

the changing pattern of sports activity and organisation, has had an important effect on the configuration of green recreational space, so has the supply, and in the next section we turn to outline the changing nature of public provision of sports areas in European cities.

Sports areas and municipal policy making

In many European countries city governments and other local authorities have been the key providers of sports and recreation areas since the late nineteenth century. What has influenced urban policy-makers to provide and maintain sports grounds and other recreational facilities? How have their motives changed over time? And to what extent did municipal sports policies get caught up with other pressing policy issues such as public health, law and order, social welfare and economic development? The following discussion focuses on the role and rationale of city governments in providing sports and recreation facilities in Western and North European countries. In Southern Europe city governments have often played a less active part, as Fulvia Grandizio's Chapter on Italian cities clearly shows.

In the decades preceding the First World War the increasing interest in improving the health of the nation and boosting national efficiency put pressure on municipal authorities to seriously reconsider their public health policies. In the nineteenth century the focus of health policies had been on improving urban sanitary conditions: water supply, sewage system and refuse disposal. By carrying through the sanitary reforms, municipal health authorities had managed to overcome devastating epidemic diseases such as cholera and typhoid, but many other serious health problems remained, including tuberculosis which killed thousands of young people in all European countries each year. After much discussion and careful consideration municipal authorities in most European countries started to shift their attention from sanitary issues to people and their habits at the turn of the twentieth century. The aim was to safeguard the health of urban populations, and especially the urban youth, and to build up their resistance to diseases by encouraging them to adopt hygienic habits and a healthy, active lifestyle.[19] To make the lifestyle changes possible, municipal authorities sought to provide health education but also to improve the access to green spaces and to provide public play grounds and sports areas where children and young people could play and exercise.[20]

---

19 For the changing focus of public health policies, see M. Niemi, *Public Health and Municipal Policy Making: Britain and Sweden 1900–1940*. Aldershot: Ashgate 2007; S. Sheard and H. Power (eds), *Body and City: Histories of Urban Public Health*. Aldershot: Ashgate 2000; D. Lupton, *Medicine as Culture: Illness, Disease and the Body in Western Societies*. London: Sage Publications 1995; M. Jauho, *Kansanterveysongelman synty: Tuberkuloosi ja terveyden hallinta Suomessa ennen toista maailmansotaa*. Helsinki: Tutkijaliitto 2007.
20 See for example, C. Nolin, 'Public parks in Gothenburg and Jönköping: secluded idylls for Swedish townsfolk', *Garden History* 32 (2004) No. 2, 197–212; D. Pomfret, *Young People and the European City: Age Relations in Nottingham and Saint-Etienne*. Aldershot: Ashgate, 2004.

In the interwar period and immediately after the Second World War, city governments intensified their efforts to promote the physical well-being and moral socialisation of the young by providing new sports and recreation opportunities. Both cities and states saw sports and recreation areas as an important element in emerging welfare policies, and after 1945, a vital part of a new welfare society. Some proponents of sports areas emphasised the value of sports as such, whereas others aimed to promote activities combining the beneficial effects of sports and 'nature'. They stressed that a harmonious combination of physical exercise and nature experiences would help maintain the well-being of city dwellers and raise them to a higher level of fulfilment in their lives.[21] The types of sports and recreation sites that were built depended on these ideals but also on negotiations between different interests and groups. In the recent volume on *The European City and Green Space: London, Stockholm, Helsinki and St Petersburg, 1850–2000* (2006), there is a wide-ranging discussion of the factors shaping planning policy on green space (from urbanisation itself to international ideas and municipal expansion); the role of planners, architects, politicians, developers and others in decision-making; and the effects of such policies on the urban landscape.[22]

Sports policies have always served many objectives, and this was particularly true in the tumultuous inter- and post-war periods. States and cities did not provide sports areas only as part of welfare policies but also to promote national political and military agendas, especially during the interwar years. Furthermore, official support for sporting activities and recreation facilities was heavily influenced by the need to enhance social cohesion and contain conflicts in urban society. Municipal authorities were (and still are) important intermediaries in the formation of collective identity. In this role, they encouraged social networks and associational life based around workplaces, schools, clubs and societies. These collectivities – from cooperative societies to drama groups and sports clubs – had an important role in modern urban society, since they provided 'a basis for orderly belonging', diminishing apathy and disaffection.[23] Furthermore, sports activities and sports clubs were seen as an efficient way to discipline people and to channel their aggressions into socially acceptable avenues. In pre- and post-war cities, where different social, political and language groups contended for power and influence, sports activities provided a safe outlet for tensions.[24] However, the solution to one problem often created other problems. As discussed by Katri Lento and Suvi Talja in Chapters Two and Three, the fact that sports policies were seen as a means of containing urban conflicts perpetuated the existing gender inequality in the public provision of sports facilities. It was usually young men who were seen as a threat to public order and were to be

---

21 Pomfret, *Young People and the European City*; D. Pomfret, 'The "city of evil" and the "great outdoors": the modern health movement and the urban young, 1918–1940', *Urban History* 28 (2001), 405–427.
22 Clark (ed.), *The European City*.
23 M. Savage and A. Warde, *Urban Sociology, Capitalism and Modernity*. London: Macmillan 1993, 150–151.
24 See for example, Aalto, 'Urheilu – viihteestä viihdeteollisuudeksi'.

kept out of the streets, and so most publicly funded sports grounds were primarily designed for them.

In the late twentieth and early twenty-first centuries, the responsibility for the provision and management of sports areas has shifted, to an extent, from city governments to other agencies and the role of sports in containing urban conflicts has clearly decreased. However, sports and recreation areas can still be high on local as well as national and international agendas. In recent years, numerous campaigns have been launched to widen participation in sport, for example, to tackle obesity, which is now accepted as a major public health problem in Europe, or to enhance the quality of life of the European aging population. Motivated by these new challenges, city governments in many European countries have made efforts to cater for a wider range of city dwellers, including women, elderly people and growing ethnic minorities, and for different sporting and recreation interests.

The provision of sports areas and other recreational facilities have also figured prominently in the marketing activities of cities in recent years. In responding to challenges posed by globalisation and intensified inter-urban competition, many cities have sought to attract new residents by portraying themselves not only as good places to do business but also as good places to live.[25] Glossy city brochures promise new residents good quality living environments with a wide range of recreational and sporting opportunities in a safe and attractive environment. Furthermore, numerous cities across Europe are interested in hosting international sporting events which are seen as an efficient way to boost both the local and national economy and to improve the image and visibility of the city. Even though the high hopes placed on the Olympic Games and other major sporting competitions are not always realised, these events can profoundly change the structure and image of cities, as shown by the examples of Amsterdam in 1928 (Chapter Four by Pim Kooij), Helsinki in 1952 (Chapter Two by Katri Lento) or Athens in 2004.[26] Such events can also have implications for the urban environment.

Sports areas and the urban environment

The ecological significance of sports and recreation areas in European cities is increasingly recognised but so far research on the subject has tended to be patchy and selective. Certainly sports areas comprise a significant part of the urban landscape, and typically cover large areas. For instance, an 18-hole golf course occupies over 50 hectares on land. As already noted, in

---

25 M. Niemi, 'Making a comeback: the politics of economic development in Pittsburgh, Sheffield and Tampere', *Jahrbuch für Wirtschaftsgeschichte* (2000) II, 71–84; P. Hubbard and T. Hall. 'The entrepreneurial city and the "new urban politics"', in T. Hall and P. Hubbard (eds), *The Entrepreneurial City: Geographies of Politics, Regime and Representation*. Chichester: John Wiley 1998, 1–23.
26 See also M. Roche, *Mega-events and Modernity: Olympics and Expos in the Growth of Global Culture*. London: Routledge 2000.

some cities sports areas are more extensive than nature reserves, and consequently, are a major ecological resource in cities. However, public opinion has often seen sports areas as harmful to the urban environment and to biodiversity. Playing fields, in particular, are frequently regarded as poor environments for wildlife because of the need for regular mowing and other management activities. However, even these 'green deserts' can support a variety of species, and formal parks (and amenity open space) tend to be rich in biodiversity because of their considerable degree of structural diversity.[27]

Historically, many sports areas have been created in areas of open space that once supported semi-natural habitat, and this habitat regularly survives within the sports area. Such relic features offer habitat for species and can provide the resource from which more extensive areas of grassland or woodland may be restored or created. Where there are no remnants of former habitats, landscaping techniques may be applied to create new habitats (such as ponds or wildflower meadows). Alternatively, maintenance regimes, which became increasingly stringent in the late twentieth century, can be amended to allow greater structural diversity. Relaxing mowing regimes, cutting hedges less frequently or delaying the removal of accumulated leaf litter are some options. Many of these measures can be used in urban sports areas. Overall, open sports areas are important for creating habitat for species and maintaining ecological values in cities. In particular, there is growing evidence, documented in Chapters Eight and Ten below, that golf courses are contributing to the biological diversity of a city.[28] Golf course landscapes typically contain diverse habitats, ranging from ponds to streams, wetlands to grasslands, and woodlands to mature forests. This variety of habitats provides unique opportunities for wildlife.[29]

Arguably, several factors have affected biodiversity in open sports spaces. One is a positive relationship between species diversity and the size and heterogeneity of green areas. In addition, there is a positive correlation between the age of a green area and its species richness (see for instance, the discussion by Suvi Talja in Chapter Nine on the richness of vegetation at Helsinki's Tali golf course, opened before the First World War) Other variables, such as the connectivity between green areas, may also positively affect their biodiversity, but this issue remains the topic of debate amongst ecologists. Intensity of management and use are also likely to affect species diversity.[30] However, little ecological research has been done specifically on sports areas to unravel the mechanisms governing the biodiversity of

---

27 S. Carruthers, J. Smart, T. Langton and J. Bellamy, *Open space in London: Habitat handbook 2*. GLC. 1996.
28 R.A. Tanner and A.C. Gange, 'Effects of golf courses on local biodiversity', *Landscape and Urban Planning* 71 (2004), 137–146.
29 M.J. Santiago and A.D. Rodewald, Considering wildlife in golf course management. Ohio state university extension fact sheet. 2004.
30 A-C. Grandchamp, J. Niemelä and J. Kotze, 'The effects of trampling on assemblages of ground beetles (Coleoptera, Carabidae) in urban forests in Helsinki, Finland', *Urban Ecosystems*, 4 (2000), 321–332.

such areas. Here two chapters (Eight and Ten) shed important new light on this issue.

In European cities sports areas may present a significant opportunity for consciousness-raising about the environment. Far more people are likely to visit their local park or playing field than their local nature reserve. Providing information about the biodiversity of the local park is the first step in promoting a greater appreciation of biodiversity in general. While public opinion may view sports areas as harmful to the urban environment and to biodiversity, urban sports areas may host a diverse set of habitats for numerous species, as we see in this book. Golf courses have been shown to be at least as diverse as many natural landscapes. With sympathetic management and maintenance, their ecological value may be enhanced. Thus golf course landscapes can be enriched ecologically by increasing their habitat heterogeneity. Transforming a golf course into a wildlife sanctuary is also an opportunity both to improve the quality of the course for the players, and to foster positive community relations, as shown in this book. Every golf course has the chance to make a significant contribution to the environment.

The high ecological values of urban sports grounds highlight the importance of innovative planning approaches and methods. In recent years urban planners have begun to find ways to preserve biodiversity as cities expand outward and subsequently modify the natural habitat. Planning sports grounds so that they enhance the maintenance of biodiversity is a way to ameliorate the detrimental effects of urban expansion. However, none of the efforts should be applied without proper knowledge of ecological principles and urban ecosystems. This is where the study of urban ecology can show its potential.

Individual Chapters

In the first part of the book the case study chapters shed important light on the interplay of international influences and municipal/national politics in the provision of sports areas and other recreational facilities. In Chapter Two Katri Lento analyses the process whereby the sports movement, which spread across Europe in the late nineteenth and early twentieth centuries, became integrated into the Finnish political and social structures. During the period of Russification around 1900 and in the first decades of Independence after 1917, many Finns found in sports a powerful source of identification with the nation, but at the same time sports structures and organisations as well as sports spaces reflected and reinforced divisions associated with class, language and gender. Sporting activities and sport related spaces could – as Lento clearly shows – simultaneously enhance people's common national identity and their separate class, language and gender consciousness. The same theme – the interaction of sports and politics – is pursued in Chapter Three by Suvi Talja. She analyses the tensions between the ideal and real in the provision of sports areas and other recreational facilities in Helsinki from the 1960s to the present day. In Helsinki, as in Finland generally, the authorities have been committed to the 'sport for all' principle, but in reality both municipal and state sports policies have favoured competitive and top-level sports over informal sports, and 'masculine' sports over 'feminine' sports. Furthermore, there have been considerable variations in the provision of sports areas and other recreational facilities in different districts of Helsinki.

For the Netherlands Pim Kooij (Chapter Four) explains the strong development of sports from the late nineteenth century with the take off of mass sports in the inter-war era, encouraged by some of the general factors mentioned above, but also in part by the distinctively Dutch policy of 'pillarisation'. He shows how sports grounds were constructed as part of old parks and in newly established parks, were incorporated in the industrial settlements, and were created as specialist sports facilities. At the same time, he also demonstrates the significant variations in developments between four Dutch cities – emphasising the influence of local municipal and other factors. In the case of Germany (Chapter Five) Christiane Eisenberg describes both the early development of sports and sports areas before the First World War, and the large-scale expansion of facilities after the war linked to the mass popularisation of sport. At the same time, she emphasises that most pitches were made from ash or asphalt and that indoor facilities were common. After the Second World War she highlights the way that the Cold War tension between East and West Germany promoted the growth of international competitive sports and construction of large sports complexes. The decline of popular support for organised sports from the 1980s has led to a 'greening' of sports areas with informal sports often taking place in urban woodlands and with a proliferation of commercial golf courses on the periphery of German towns.

In Chapter Six Fulvia Grandizio examines the general failure of municipal provision of sports and recreation grounds in the Italian city of Turin, which she attributed to the ineffectiveness of urban planning and the power of private developers and other interest groups. The most important growth of facilities came through the paternalist initiative of Fiat and other industrial companies concerned to improve the health and productivity of their workforces. After the Second World War urban demand ran far ahead of increased provision of sports. In recent years the most significant increase of sports sites has sprung from the flowering of largely private golf courses in the Turin region, once again mainly catering for the rich.

In the second part of the book the focus shifts to the relationship between sporting activities and the urban environment. Several Chapters look at the social and ecological significance of sports areas in cities and examine the building and maintenance of sports areas in the wider context of urban planning. Jürgen Breuste addresses these questions in Chapter Seven by looking at the development and changes in planning of urban green space in Germany in the late twentieth century. In particular, he describes how ecological issues have been taken into account during the different eras of planning. The destruction of many German cities during the Second World War, and the country's political division after the war played an important role in post-war urban development. The wastelands created by the war contributed to the development of research on urban ecosystems and their management. In the 1970s, urban ecology developed strongly and rapidly, particularly in Berlin, influencing the discipline in Europe and worldwide. According to Breuste, this surge in research and the growing public awareness of urban environmental problems in the 1970s stimulated an increased interest among urban planners towards urban nature. The ecological significance of green spaces influenced urban green space management from the 1970s to the end of the 1990s.

In Chapter Eight Jarmo Saarikivi explains how golf courses, if planned and managed properly, can contribute to the ecological diversity of an urban area. He notes that worldwide, there are over 30,000 golf courses and 55 million people play golf. Many countries are experiencing a golf boom and new courses are being built at a rate that makes golf course development one of the fastest growing types of land development in the world. He explains that this development has environmental impacts, and the public perception is that golf courses are detrimental to the environment. However, as the study points out, golf courses can also have a positive effect on urban biodiversity. Many golf courses promote nature conservation and host rare plant and animal species.

Suvi Talja looks at the history of Tali golf course, the oldest and best known golf course in Finland, in Chapter Nine. The Tali golf course was established in 1932 outside the municipal borders of Helsinki. As the city grew in the post-war decades both in population and area, the golf course which was initially set in the middle of agricultural fields became a contested urban space where different groups – players, politicians, and city-dwellers – continuously negotiated their position. Finally, Talja investigates the process by which the golf course area where different interests clashed in the 1960s and the early 1970s gradually turned into an area where different interests came together and found common ground.

In Chapter Ten Philip James and Emma L. Gardner present a British case study of urban golf courses and their contribution to the maintenance of biodiversity. Their exploration is based on a case study of a new golf course constructed in the 1990s in Salford. Their analysis indicates that many of the features of the wider landscape were incorporated into the golf course, and this leads on to a new model for the inclusion of sports space, and in particular golf courses, within urban ecological frameworks. One outcome of this model is a 'toolkit' recommending the best ways of improving urban areas for biodiversity, together with spatial information about where the tools in the kit should be put to best use. They conclude that sports fields, such as golf courses, are important green spaces for the maintenance of biodiversity and quality of life for residents.

In the Federal Republic of Germany, as in Finland, golf was for much of the post-war era the domain of wealthy and influential people, while in the German Democratic Republic the sport was heavily repressed for ideological and other reasons. The golf course building boom did not begin in Germany until the late 1980s, and even then the growth of golf was slower than expected. By looking at the development of golf in Germany in general and in Berlin and Brandenburg in particular, Christiane Eisenberg and Reet Tamme (Chapter Eleven) reveal the important role that nature protection interests have played in regulating and restraining the golf boom in Germany. By the late 1980s, when the golf boom began, nature protection movements and environmental organisations had already managed to develop a high level of sensitivity to possible environmental problems associated with building projects. Consequently, local authorities scrutinised golf projects for their effects on the environment and applied standards that were often too high for golf course investors. However, as Eisenberg and Tamme show, the *German Golf Association* (DGV) rose to the challenge. Once the DGV acknowledged that environmental standards had come to stay and that further resistance would be futile, they sought to use environmental approaches and arguments to their own benefit by promoting golf 'as the quintessential agent of nature protection'.

In Chapter Twelve Niko Lipsanen takes a different approach to examine the relationship between sporting activities and the environment. He looks at the changes in the layout, vegetation and landscape features in sports sites in Helsinki by comparing two sets of photographs; the first set taken in 1976 and a second taken by Lipsanen himself in 2006. At first sight changes seem to be small, and many sports sites in Helsinki look much the same as they did 30 years earlier. Some changes, however, are significant. For example, sports sites have become more exclusively reserved for sporting activities. Instead of promoting the multifunctionality of sports spaces, the municipal authorities have – wittingly or unwittingly – restricted the use of sports sites to sporting activities by building fences and letting vegetation grow. Such exclusion strategies have many negative effects but – as Niko Lipsanen argues – the clear borders may also increase the 'placeness', the place identity, of the sports sites for those who are using them.

In conclusion, the development of sports spaces in European cities and their impact is clearly an important and complex topic, with significant national and local variations and relevance for current urban policy, ideas about political ecology, and the wider environmental debate. We hope that the following case studies and concluding chapter will put the subject firmly on the scientific agenda and open up further discussion from a comparative and transdisciplinary perspective.

PART I:
National Trends in Sports Areas Development

PART I.

KATRI LENTO

2 A Question of Gender, Class and Politics: The Use and Provision of Sports Grounds in Helsinki c.1880s–1960s

This chapter examines the development of sports areas in modern Helsinki up to the post-war era. It raises questions about how the building of sports sites influenced the urban structure and how sport affected the use of green space in Helsinki. It also considers issues about who provided sports places, whether it was the municipality or other actors, as well as for whom sports places were planned and built. A fundamental concern of this chapter is the question of access, availability and control in regard with urban space and sport. Here it is argued that sports activity in Helsinki was often a question of politics, class and gender, and that sports grounds were more often than not highly contested urban space.

Healthy bodies in healthy cities

In the growing industrialised cities an increasing concern about the unhealthiness of urban life was voiced during the nineteenth century. For social reformers healthier cities meant poor people living in better housing and preferably in a 'green' environment; thus closer to 'nature'. The new and growing middle class was also starting to long to move into a greener environment. Several privately financed garden communities were founded in Helsinki and the surrounding area at the beginning of the twentieth century.[1] These garden suburbs, like the early nineteenth century parks in Helsinki, were financed and supported by better-off private citizens.[2] Increasingly there was the view that not only should the urban living environment be made greener and healthier but also the people living in cities should be fitter. With the ideology of healthy urban space came that of the healthy body. Physical fitness was connected to mental fitness, or, as Ivar

---

1   These were based on the planning ideas of the international Garden City Movement spread to Finland by the end of the nineteenth century and adopted by progressive minded architects, many of whom themselves moved into the new garden suburbs such as the Kulosaari and Kauniainen garden communities.
2   In Helsinki's case it was the wealthy merchant Henrik Borgström, who was the most important park advocate in the early nineteenth-century century.

Wilskman,[3] an important figure in the Finnish sports movement, wrote in 1896: 'in a healthy and able body lives a sober, decent and free spirit'.[4]

An important change in Helsinki came when the provision and care of green spaces was incorporated into municipal policy. The first municipal policy on parks was established in 1889. It meant that the maintenance of parks was taken over by the city instead of being left to private individuals who had made parks into profitable enterprises with, for instance, cafes, restaurants and other amusements. Initially, aesthetic values had priority in the municipal design of parks but gradually playgrounds and sports facilities were introduced into Helsinki's parks from the 1890s.[5] Reformist ideas on healthy urban space created a new type of park that started to dominate park planning in Finnish cities from the turn of the twentieth century. The earlier park ideal, the scenic park, where the aim had been aesthetic pleasure, was replaced by the reform park. A significant difference between these two types of parks featured in the way they were to be used. Recreation in the new reform parks was meant to be active. According to Maunu Häyrynen, the underlying aim of the municipality with the new reform parks was to socialise the young and working class people by offering regulated exercises and supervised play, while at the same time making the inhabitants healthier.[6] This also meant strict control of 'proper' park behaviour.

Debate over the healthiness of the urban environment was active from the mid nineteenth century. The role and involvement of the municipality was one of the main questions. A good example of the demands made on the city of Helsinki was the address the physician Tor Waenerberg made in 1914 to the City Council on behalf of the *Finnish Association for Sport Teachers*. Waenerberg spoke of the dire living conditions of the poor people in Helsinki and emphasised the municipal obligation for providing healthy public places, especially for children: 'the city has an obligation to offer all small children living in those airless cupboards a chance to spend time outside for as long as possible during the day time'.[7] The role of the municipality as provider of recreational activities as well as sports sites was also stressed by others, like the school counsellor Erik Mandelin writing in 1919 in the

---

3 Ivar Wilskman (1854–1932) founded many of the first sports clubs in Finland and was one of the founders of *Finnish Athletics Association* (SVUL) and its first president. He was actively involved in Finnish sports education, both at the university and in secondary schools. He founded the first Finnish language sports club in Helsinki in 1878, *Helsingin Alkeisopiston Turnarit*, and the first Finnish language sports magazine, *Urheilulehti*, and wrote several textbooks on sport and sports education. In the 1920s and 1930s he became known for his ideas on eugenics and the Finnish race.
4 I. Wilskman, *Voimistelun käsikirja kansa- ja alkeiskouluja varten*. Helsinki: G-W. Edlund 1906: 'terveessä ja työkykyisessä ruumiissa asuisi raitis, siveellinen ja vapaa henki'.
5 M. Häyrynen, *Maisemapuistosta reformipuistoon: Helsingin kaupunkipuistot ja puistopolitiikka 1880-luvulta 1930-luvulle*. Helsinki: Helsinki–Seura 1994, 237.
6 M. Häyrynen, 'Vihreät vuosirenkaat: Helsingin kaupunkipuistojen kehitys', in S. Laakkonen et al. (eds), *Näkökulmia Helsingin ympäristöhistoriaan: Kaupungin ja ympäristön muutos 1800- ja 1900-luvuilla*. Helsinki: Helsingin kaupungin tietokeskus, Edita 2001, 41.
7 Helsingin kaupunginvaltuuston pöytäkirjat (Helsinki City Council Records) 1914/no. 50, 18: 'Sen tähden onkin kaupungin välttämätön velvollisuus hankkia kaikille noille ummehtuneiden. liika-asuttujen komerojen pienokaisille tilaisuutta oleskelemaan ulkosalla niin suuren osan päivästä kuin mahdollista'.

Figure 2.1 Children playing in a sand pit in Helsinki in 1933. Municipal play parks were seen as important in offering children opportunities for fresh air and exercise. (Photograph: Toini Saloranta, Albumit auki).

municipal journal *Suomen Kunnallislehti*: 'park allotments, playgrounds, school gardens, sports fields, summer colonies, – different activities for different age groups, which means that the task of taking care of the healthy development of children from the beginning belongs to the municipality'.[8]

Park reform was only part of the growing interest of municipal government in social issues. The change in the municipal attitude towards urban green space is shown by the way that green areas started to be included in urban planning from the 1910s. In 1900 the amount of urban park land had been 60 hectares, and in 1930 it had already increased to 140 hectares. By the year 1930 green areas were already a standard part of the planning system and as a result Helsinki quickly became greener.[9] The new building by-law from 1931 was important both for greening Finnish cities as well as making green space the responsibility of the municipalities. According to the new law, cities were obliged to include a certain amount of parks, gardens, play and sports fields in their city plan.[10]

---

8 *Suomen Kunnallislehti* 1919–1921, n:o 7–8, 107; 'Puistosiirtolat, leikkipaikat, koulupuutarhat, urheilukentät, kesäsiirtolat, – eri-ikäisille eri toimintoja eli kunnan tulee huolehtia lasten terveestä kehityksestä alusta asti'.
9 E. Vasara and O. Havuaho, *Puistotyöntekijäin ammattiosaston historia, KTV osasto 16, 50 vuotta 1933–1983*. Helsinki 1983, 17–19. See also K. Lento, 'The role of nature in the city: green space in Helsinki 1917–1960' in P. Clark (ed.), *The European City and Green Space: London, Stockholm, Helsinki and St Petersburg 1850–2000*. Aldershot: Ashgate 2006. The involvement by the municipality, in addition to leading to an increase in parks and recreation areas it also resulted in ambitious municipal 'green' housing projects in the districts of Vallila and Käpylä in the 1920s.
10 Building by-law, Helsinki 24 April 1931, Finland's statute book 1931, no 145.

From a upper class activity to mass sport

Sport became fashionable in Finland in the 1870s among the upper classes. Popular sports at the time included sailing, rowing, skating, horseback riding and bowling.[11] These activities did not require purposely planned sites; thus bowling and cycling were popular pastimes in Helsinki city parks. In the late nineteenth century gymnastic exhibitions were held there too. The first in Finland was held in Kaisaniemi Park in 1886 and also included a four event athletics competition.[12] Other smaller sports competitions were also held in the Helsinki parks, mostly in Kaisaniemi.[13] Another location for upper class sport in Helsinki was the surrounding sea area. Helsinki with its extensive shoreline and a small archipelago was ideal for sailing which became a popular elitist past time during the late nineteenth century. Several yacht clubs built harbours and sailing pavilions on the islands surrounding Helsinki.

Important prerequisites for exercising sport include having the free time, as well as financial means to do so. Many of the sports that the affluent classes patronised in the nineteenth century were time-consuming, as well as expensive to pursue; good examples being sailing and horse riding. However, this situation in which only a small part of the population were active in sport, changed around the turn of the twentieth century. A wide enthusiasm towards sport and exercise swept across Europe and spread also to Finland. Sports activity became common at all social levels, not only among the wealthier population. This generated a need for larger sports fields to accommodate the growing number of athletics and team games.[14] In Helsinki, there was a notable increase in sports societies. In 1875 there were five sports societies in the city, in 1900 altogether 26 and by 1909 there were fifty of them.[15] Following the European trend in sport, outdoor sports such as hiking became fashionable in Finland at the beginning of the twentieth century. The first sports clubs were often privately founded, for example teachers were actively involved in founding sports clubs, and some were organised by workers' societies or some professional guild.[16] As sport became more popular they started to receive more recognition and funding from the state and the municipality.

After Finnish independence salaries rose and the working day decreased to eight hours. This encouraged a marked increase in recreational activities. Shorter working days also meant a rapid growth of trips to the outer forest

---

11 Sailing was the first sport in Finland to be organised in societies. The first sailing club was founded in the town of Pori in 1856.
12 T. Aalto, *Helsingin kaupungin urheilu ja ulkoiluhallinto 1919–1969*. Helsinki: Helsingin kaupungin julkaisuja N:o 22, 1969, 13.
13 For example a cycling club held a competition in the Kaisaniemi park on a Sunday morning in October 1887. There were altogether seven competitors who competed in 10 000 and 5000 meters. All of them were men as it was not customary for women to compete in sport at the time. *Wiipurin Sanomat* 5.10.1887.
14 Häyrynen, 'Vihreät vuosirenkaat', 43–45.
15 Aalto, *Helsingin kaupungin urheilu ja ulkoiluhallinto*, 9.
16 Aalto, *Helsingin kaupungin urheilu ja ulkoiluhallinto,* 12–15.

areas of Helsinki. By 1920 the use of public transport had risen over twentyfold since the first electric tramline started in 1901.[17] As the importance of leisure and recreation grew, spectator sports became the most popular pastime activity in Helsinki. As Table 2.1 shows the popularity of other organised events came far behind.

Table 2.1 Five most popular organised public events in Helsinki Jan. 1926 – Oct. 1926[18]

	Jan	Aug	Oct	Total
Organised sports events	52	44	8	124
Events by labour organisations	36	16	45	97
Dancing and masquerades	34	18	42	94
Religious events (unofficial or non-church affiliated)	46	9	37	92
Theatre	35	10	36	81

Not only spectator sports, but participatory sports activities became increasingly popular in Helsinki, and in the 1920s and 1930s the fastest growing recreational sector for the Helsinki inhabitants.[19] The growing enthusiasm for sport corresponded with the increasing municipal provision of sports sites. After the founding of the Helsinki City Office for Sport in 1919, the number of municipal sports sites more than doubled over the next twenty years. In 1920 there were 41 municipal sports places, by 1930 the number had grown to 62 and by 1940 to 116.[20] The city also acquired large un-built areas outside the city limits in the 1920s in expectation of the future growth of the city. These were offered for outdoor recreation by inhabitants. Also the state became involved in providing sporting activities. The Ministry of Education established a Hiking Division in 1920 and by 1934 there were already six hiking societies in Helsinki with approximately 500 members.[21]

Politics and sport

The rise of sport in Finland coincided with the end of the first period of Russification (1899–1905) and this meant that the Finnish sports movement

---

17 See also Lento, 'The role of nature in the city'.
18 Aalto, *Helsingin kaupungin urheilu- ja ulkoiluhallinto*, 20. The survey was done by counting the advertisements from three daily Helsinki newspapers (the main Finnish-speaking, the main Swedish-speaking and the social democrat newspaper): *Helsingin Sanomat, Hufvudstadsbladet* and *Suomen Sosialidemokraatti*.
19 J. Siipi, 'Pääkaupunkiyhteisö ja sen sosiaalipolitiikka', in R. Rosén et al. (eds), *Helsingin kaupungin historia V:1*. Helsinki: Helsingin kaupunki 1962, 137–379.
20 K. Ilmanen, *Ensimmäisenä liikkeellä: Helsingin kaupungin liikuntatoimi 1919–1994*. Helsinki: Helsingin kaupungin liikuntavirasto 1994, 69.
21 L. Lehto, *Elämää helsinkiläismetsissä: Helsingin kaupunginmetsänhoidon syntyhistoriaa*. Helsinki: Helsingin kaupunginmuseo, Memoria 1989, 66.

Figure 2.2 The Ball Park (Pallokenttä) sports field in Helsinki during a grand fête of the Finnish Athletics Association (SVUL) in 1947. (Photograph: The Sports Museum of Finland).

was politicised from the beginning, with a special focus on competitive sport. The success of Finnish athletes, such as the runner Hannes Kolehmainen who won three gold medals at the Stockholm Olympics in 1912, was important in strengthening national feelings, as well as in gaining Finland recognition as a nation in its own right. However, even if Finnish athletes competing abroad had a unifying effect at the national level, within the Finnish sports movement there were severe conflicts based on language and class questions, and by the 1910s the sports clubs were already divided into Swedish and Finnish language organisations.[22]

After the Finnish civil war in 1918, fought between the socialist 'reds' and the conservative 'whites', the division of the sports movement was unavoidable, especially as sports societies, as well as youth societies and volunteer fire brigades, had been, for both sides, places from which to recruit fighters.[23] *The Finnish Workers' Sports Federation* (TUL) was founded in

---

22 S. Karvala, 'Urheilu myös jakoi katajaista kansaa', in H. Roiko-Jokela and E. Sironen (eds), *Urheilu katsoo peiliin*. Jyväskylä: Atena Kustannus Oy 2000, 172–180.

23 S. Hentilä, 'Urheilu järjestäytyy ja politisoituu', in T. Pyykkönen (ed.), *Suomi uskoi urheiluun: Suomen urheilun ja liikunnan historia*. Helsinki: Liikuntatieteellisen seuran julkaisu, VAPK-kustannus 1992, 150.

1919, and competed with the Finnish central organisation for sport, *Finnish Athletics Association* (SVUL), which had been founded in 1906. However, the division into workers' sports clubs and 'bourgeois' sports clubs had already started at the beginning of the twentieth century. In 1909 there were 119 workers' sports clubs in Finland.[24] This division of the Finnish sports field continued until 1993.[25] The political division of sport had a profound impact on Finnish sports life. The importance of the *Workers' Sports Federation* increased rapidly during the post-war decades when a *TUL* sports club was founded in practically every village in the countryside. The division also had a profound impact on the use of urban space. Seppo Aalto argues that the politicisation of sport in Helsinki led to a politicisation of urban space as sports societies connected to different political parties used certain sports fields in specific parts of the city. In addition, grants by the city to sports clubs were given out on political grounds until 1959, enhancing the political aspect to sport.[26]

Gender and sport

Another important dimension concerning the use of sports sites is gender-related. How different was it for women and girls to play sport in comparison with men and boys and how gendered was the space reserved for sport in Helsinki? Already during the nineteenth century there were public calls for girls to have opportunities for physical exercise. For example an article in the local newspaper *Helsingin Uutiset* from 1863 demanded a special park for girls to 'play' in and stated that gymnastics was fundamental for girls' schools.[27] The 'father of Finnish sport', Ivar Wilskman, was quoted in 1898 in the journal *Suomen Urheilulehti* as follows: 'women are in need of health just as much as men, but to them physical development is almost even more important than for men'.[28] In general, however, physical exercise, especially of the strenuous kind, was seen as unfeminine and even dangerous for women's health, especially for female fertility.[29] Women were encouraged to exercise moderately to achieve and maintain physical well-being. This attitude of discouraging women from sport also meant that before the 1960s

---

24 Hentilä, 'Urheilu järjestäytyy ja politisoituu', 142.
25 In 1993 competitive and world class sports formerly under *TUL* and *SVUL* were handed over to the sports federations and to Finland's Olympic Committee to manage.
26 S. Aalto, 'Urheilu – viihteestä viihdeteollisuudeksi', in H. Schulman, P.Pulma and S. Aalto, *Helsingin kaupungin historia vuodesta 1945: 2*. Helsinki: Helsingin kaupunki 2000, 345–355.
27 'Naiskouluissa on voimistelu välttämätön ja paitsi sitä pitäisi niillä olla puisto, missä tytöt voivat leikitellä', *Helsingin Uutiset* 22.06.1863.
28 '...naisetkin jotka tarvitsevat terveyttä samassa määrin kuin miehetkin, mutta naisille on ruumiin kaikinpuolinen kehitys miltei vielä tärkeämpi kuin miehille', as quoted in K.U.Suomela, *Ivar Wilskman: Suomen urheilun isä*. Helsinki: Osakeyhtiö Valistus 1954, 101. Original quote in *Suomen Urheilulehti* n:o 1 (1898), 11.
29 J. Kari, 'Naiseuden kuperkeikka – "Hiihtoäiteen" tarinoita', in *Suomen urheiluhistoriallisen seuran vuosikirja 2000*.

sport played a minor role in the physical education of girls at school.[30] Women were discouraged from competitive sport in the early decades of the twentieth century, even though there were competitions open for women. The negative attitudes towards women competing in sport came to a conclusion in 1923 when the Finnish central organisation for sport, *SVUL*, ended women's regional competitions in athletics for some years. In comparison, non-competitive forms of exercise such as gymnastics and dance were regarded as suitably feminine for girls and women. The opposition against women competing in sport became less pronounced as Finnish women started being successful in international games in the 1930s which meant more resources for the training of women athletes.[31]

According to the sports historian Leena Laine, many of the early pioneers of sports education in Finland created physical education for women and girls to meet specifically 'female needs' and thus wilfully feminised certain types of sport.[32] One of the most active advocates of female sport in the early years was Anni Collan[33] (1876–1962), a gymnastics teacher and inspector of female sports education for the National Board of Education. Although Collan saw the advantages of sport and physical exercise for women, offering them an escape from the dreariness of domesticity, she also believed that female exercise should be strictly regulated and limited. Collan believed in a scientific approach to gymnastics and stressed the importance of good physical condition for women; not least, so that women would be better able to bear and raise children. Collan also saw a connection between exercise and the need for outdoor recreational areas. During her travels in Northern America in 1916, she had been introduced to the American 'playground movement', and after returning to Helsinki, Collan spoke actively on behalf of the need for childrens' playgrounds and better school sports grounds. In her writings in several journals she often quoted their slogan: 'A playground built today will make tomorrow's prison and hospital unnecessary'.[34]

The gender division in sport resulted in a spatial division of urban space, as in the case of sport and politics. One difference between so-called feminine

---

30 H. Meinander, 'Koululiikunta etsii paikkaansa', in Pyykkönen (ed.), *Suomi uskoi urheiluun*, 295.
31 L. Laine, 'Ruumiinharjoitusten monet muodot', in Pyykkönen (ed.), *Suomi uskoi urheiluun*, 205–206. The new interest and support women athletes started to gain in the 1930s waned as the success of Finnish women did not measure up with European and American women who had a couple of decades head start in professional sport training. Even as late as in the 1950s Finnish women skiers, even successful ones, were not seen as valuable athletes. This comes clear from the way they were treated by the Finnish Ski Federation as well as by their local sports clubs. They were not provided with personal trainers, nor did they get the same financial rewards from the Federation as men. Their medals, even Olympic ones, were not officially celebrated. Kari, 'Naiseuden kuperkeikka'.
32 L. Laine, 'Naisten urheilun reunaehdoista: kokemuksia yleisurheilusta ja hiihdosta 1940-luvulle', in K. Ilmanen (ed.), *Pelit ja kentät: Kirjoituksia liikunnasta ja urheilusta*. Jyväskylä: Jyväskylän yliopisto: Liikunnan sosiaalitieteiden laitos 2004, 119.
33 Anni Collan was also active in the temperance and Fennoman movements and involved in the founding of the Girl Scouts organisation in Finland. Collan herself was the perfect representative of the new modern women: single, career orientated and active in society.
34 E. Vainio, 'Anni Collan ja naisliikunnan mahdollisuus', in *Suomen urheiluhistoriallisen seuran vuosikirja* 1999, 159–168; Meinander, 'Koululiikunta etsii paikkaansa', 283–284.

Figure 2.3 Two female gymnacists by a lake in Tampere c. 1950. According to a commonly held view, female exercise was meant to be more an aesthetic experience for the viewer, rather than strenuous physical activity for women. (Photograph: Jussi Kangas, Vapriikki Photo Archives).

exercise, gymnastics and dance, and sport more commonly exercised by men and boys, was that the former were mostly practised indoors and away from the public eye. What made female exercise visible and brought it out into the open were gymnastics competitions or group performances in gymnastics or dance. These took place in various sports events or public festivals. At the same time, women exercising was meant principally to be an aesthetic experience for the spectators rather than a physical activity for the participants. Popular sports for boys and men, on the other hand, such as Finnish baseball, football and athletics were exercised in the open outside. In consequence outdoor sports sites in Helsinki were heavily gendered space, and mainly used by men and boys.

These elements of segregation and exclusion with regard to female exercise can be observed in the plan for a sports park in Eläintarha by the city gardener Svante Olsson in 1914.[35] Olsson's plan was heavily influenced by Bertel Jung's earlier proposal for a Central Park in 1911 as well by a study trip Olsson had made to Germany.[36] The plan included a special sports ground for women that was to be separated from the sports grounds for men. To further emphasise the difference between male and female sport, next to the

---

35 S. Olsson, 'Förslag till en idrottspark i Djurgården i Helsingfors', *Finsk trädgårdsodlaren* 6 (1914), 111–116.
36 Häyrynen, 'Vihreät vuosirenkaat', 45.

women's sports ground a place for children's playgrounds and a paddling pool was set aside, thus allowing women to take care of their children during exercise. This idea of giving mothers the possibility of looking after their children while performing other tasks was also found in other plans of the time, influenced by new ideas of rationalising housework. Thus in Sweden, the garden designer Ester Claesson in 1925 designed a large yard area for seven houses. In the central yard there were sandpits and play areas for children with the idea that the women living in the houses could keep an eye on their children while at the same time performing household chores.[37] Olsson's plans for a sports field for women remained un-realised but the idea of women's sports grounds was brought up again in the 1930s with two unsuccessful petitions made to the Helsinki City Council.[38]

A gradual change towards the expansion of the female realm and women becoming more active outside the home is illustrated by the changing role of women in sport. This can be seen, for instance, in *Elanto*, the magazine of the co-operative Elanto, where the number of advertisements for sports equipment for women, such as skis, bicycles and warm sports clothes for winter sport increased from the 1930s.[39] The new modern urban culture of the 1920s and 1930s encouraged women, particularly educated and better-off ones, to extend the traditional boundaries of female exercise beyond traditional gymnastics. Competitive sport gained popularity and in swimming and athletics Finnish women even took part in the Olympic Games.[40] Here sport mirrored the widening freedom of choice for women in the inter-war period and their greater ability to control their free time.

Sport and the education system

How far did the expansion of the schooling system in Finland before and after independence help to erode distinctions in sports activity? In the 1872 school reform, it was made obligatory for municipalities to hand over land free of charge for school buildings and outdoor recreation activities during school time. By the end of the nineteenth century new secondary school architecture had been created. The new type of school buildings included modern gymnasiums.[41] Physical education became part of the syllabus in schools, even in girls' schools, as early as in 1872. However, as physical education for girls meant non-competitive forms of sport such as gymnastics, it was boys' schools that took the lead in introducing sport systematically into their curricula. There was a rapid increase in the number of school

---

37 C. Nolin, 'Stockholm's urban parks, 1860–1930', in Clark (ed.), *The European City*, 123.
38 Liikuntaviraston arkisto (The City of Helsinki Sports Department Archives), Naisten urheilukenttäkysymys, Anomus naisten urheilukentän perustamisesta Helsinkiin 28.8.1933 ja 7.1.1935, Ulk 25.10.1934.
39 E. Packalén, 'Naiset Elantolaisessa julkisuudessa 1920- ja 1930-luvuilla', in M. Sarantola-Weiss et al. (eds), *Kulman takana Elanto*. Helsinki: Helsingin kaupunginmuseo 2005, 84.
40 Laine, 'Ruumiinharjoitusten monet muodot', 188.
41 H. Meinander, 'Warpaille y-lös, kyykkyyn alas', in Pyykkönen (ed.), *Suomi uskoi urheiluun*, 87–89.

sports grounds in the 1920s and 1930s as sport and outdoor activities became important in the curricula of boys' schools. School sports grounds, especially in secondary schools, had a key role in the provision of sports areas in Helsinki.[42] Teachers were also influential in promoting sport during the first decades of the twentieth century. Their initiatives in the Helsinki City Council resulted, for example, in the building of new sports fields in 1916 and 1917.[43]

Henrik Meinander has argued that sport in Finland during the early years was an urban phenomenon and the education system in Finland was the main means of implementing the ideology of sports exercise. According to Meinander, sport and exercise were linked to a bourgeois lifestyle, which through an increase in external discipline, cleanliness and health as well as controlling and directing a pupils' (boys) natural need for exercise, sought to create 'new humans', who would physically act according to the needs of modern urban culture. What made sport an urban phenomenon in the beginning, was that sports education was of higher quality in the cities than in the countryside, as the schools in bigger cities had the best facilities and equipment. Meinander also underlines the difference in attitudes towards sport between city and the countryside. The population in the countryside did not appreciate the value of exercise or sport for their children as much as townspeople, since daily life at the farm provided enough exercise, and bourgeois recreational values did little to increase the crop or milk production.[44] As a new urban culture slowly emerged in Finnish cities in the 1920s, motor sport and aviation became seen as fashionable as well as distinctively urban and continental.[45]

Hosting the Summer Olympics 1952

Preparations for the Helsinki Olympic Games, first for the cancelled ones of 1940, and then for those in 1952, greatly increased the number of sports buildings in Helsinki and larger sporting facilities were planned and constructed. The first steps for organising major international sports events in Helsinki had been taken already in 1928 when the city together with a number of Finland's major sports organisations founded the *Stadion-säätiö* (the Stadium foundation) in order to create a world-class sports arena in Helsinki, with the hope of playing host to the Olympic Games some day.[46] Preparing for the Olympics changed the external appearance of Helsinki. First of all there were several new Olympic-size sports sites and facilities.[47]

---

42 Meinander, 'Koululiikunta etsii paikkaansa', 283–284.
43 Aalto, *Helsingin kaupungin urheilu- ja ulkoiluhallinto*, 15–16.
44 Meinander, 'Warpaille y-lös, kyykkyyn alas', 98–99.
45 Laine, 'Ruumiinharjoitusten monet muodot', 185–186.
46 M. Härö 'Olympiakisojen suorituspaikat', in M. Härö et al. (eds), *Olympiakaupunki Helsinki 1952*. Helsinki: Helsingin kaupunginmuseo, Memoria 1992, 14.
47 The new sport facilities built either for the 1940 Olympics or the 1952 Olympics in Helsinki were the Olympic stadium, the swimming stadium in Eläintarha, the Velodrom in Käpylä sports park, Meilahti rowing stadium, the indoor riding hall at the Laakso stadium, Messuhalli and Tennispalatsi. Extensions and restorations of the existing sports sites included the Töölö Pallokenttä for football as well as the Tali and Laakso horse riding stadiums.

Figure 2.4 The Olympic rowing stadium, Helsinki, built in 1939. The sports facilities and other buildings built for the Olympics transformed Helsinki's cityscape. (Photograph: Aarne Pietinen, Helsinki City Museum).

Most importantly, Helsinki at last gained the sports park for which the city gardener Svante Olsson had made plans in 1914. The new sports park was centred on the stadium and was connected to the Central Park. In anticipation of the Olympics the city of Helsinki also bought up privately owned sports sites such as the Ball Park (Pallokenttä) built in 1915, and thus strengthened the municipal ownership of sports sites in Helsinki.

Sports sites were not the only new additions to Helsinki's cityscape brought about by the Olympics. Two new housing projects[48], several new restaurants, outdoor cafés, an amusement park and hotels, as well as Helsinki's first camping sites were built. Building for the Olympics brought international urban culture to Helsinki. In addition to all the new cafes and restaurants, even if some were only built to last the Olympics, there was the new hotel *Kalastajatorppa*. The seaside hotel aimed at a continental, country house style with tennis lawns, parks and a waterfront cafe.[49] However, the main

---

48 The first Olympic village to house the athletes during the games, Olympiakylä, was built in 1939 and the second one for the 1952 summer Olympics, named Kisakylä, in 1951. They are both located in Helsinki's first municipally built garden suburb district Käpylä. Käpylä was the show case of municipal housing production in Helsinki in the 1930s. Both Olympiakylä and Kisakylä were marketed as examples of the high standard of Finnish architecture and planning and of its close-to-nature character.

49 Härö, 'Olympiakisojen suorituspaikat', 109–110. Not only did public space undergo a facelift, it was also hoped that private homes would be improved for the games. Marttaliitto, the Martha Organisation, founded in 1899 'to promote the quality and standard of life in the home' gave instructions on how to beautify the home environment so as to present a flattering image of Finnish homes. *Kotiliesi* 1939, 402–403 and 486–487.

impact of the Olympics was probably on a symbolic level. First of all, the choice of Helsinki to host the Olympics was of immense significance for the war-ridden nation. The 1952 Olympics became a national project with volunteers from all over the country participating in the organisation of the Games. The Games had a strong unifying effect among Finns and enhanced post-war national identity.

How far did the Olympics erode gender and class divisions in Finnish sports life? The Olympic Games in general, like other international sporting events, helped the cause of female competitive sport. In Finland the central sports organisations started to encourage women's participation in competitive sport when women started to have their own competitions in international games. In this way the Olympic Games did help erode gender divisions in sport, at least among competitive and elite sport. Along with the strong national sentiments the Helsinki Olympics evoked among Finns, spectators and on-lookers, practically the whole nation, regardless of class and gender, became emotionally involved and shared a common interest. On the other hand, the huge financial effort made for the preparations of the Olympics, especially so soon after the war years, prevented or postponed for several years the building of many smaller scale sports sites in Helsinki.[50] In this regard the 1952 Olympics strengthened another aspect of exclusiveness in sport. Instead of the class and gender dimension of the nineteenth and early twentieth century, the emphasis on competitive sport discriminated against those interested in more informal or individual recreational sport.

Sport in post-war Helsinki

In 1952 the architect Yrjö Lindegren wrote in *Arkkitehti,* the leading Finnish journal for architects and planners: 'a sports area with its extensions will develop into an oasis in society to which young people will hurry to relax, to achieve physical fitness and an active mind'.[51] During the years after the Second World War criminality and social unrest grew in Helsinki, and, as earlier, one official solution was to offer citizens, mainly young people, healthy ways to spend their leisure time. The growth in the number of municipal sports sites that had started in the 1920s continued after the Second World War.[52] According to Seppo Aalto, the main underlying reason for the city of Helsinki to grant funds for sports clubs and for building sports

---

50 Aalto, 'Urheilu – viihteestä viihdeteollisuudeksi', 388.
51 '...urheilualue laajennuksineen muodostuu yhdyskunnassa sellaiseksi keitaaksi, jonne nuoriso kiiruhtaa rentoutuakseen, hankkimaan fyysillistä pohjakuntoa ja reipasta mieltä'. *Arkkitehti* (1952).
52 For example in winter 1950 in Helsinki (population 368 519) there were 29 municipal skating rinks, 45 municipal sledge slopes and 5 municipal street sledge slopes. Skating rings maintained by sports clubs were visited by 102 622 school children. In summer 1950 the beach in Hietaranta was visited by 225 000, Seurasaari by 250 000, Pihlajasaari by 31 000 and Korkeasaari by 4000 persons. 'Helsinki tilastoina 1800-luvulta nykypäivään', Helsinki: Helsingin kaupungin tietokeskus 2000:15 tilastoja, 236.

sites, was not to guarantee the success of world class sportsmen and women, or to enhance the international sporting fame of Helsinki, or the Finnish nation. Rather the municipality was principally motivated by the need to alleviate social problems by redirecting young people into sports clubs. Aalto sees a strong element of control in the willingness of the municipality to grant funds for sport as sports grounds and halls were easily controlled space. To increase control those sports clubs receiving funding from the city were also obliged to report to the Council on their activities.[53]

As competitive sport and sports clubs were in the hands of the two central and highly politicised organisations, *SVUL* and *TUL*, the municipality concentrated on mobilising the masses. This corresponded well with the larger post-war social ideals and modernist planning theories, as well as with the focus on the creation of the welfare state. Finnish policy-makers were largely influenced by the social planning ideals of Sweden, where for example a progressive parks policy was adopted in the 1940s and 1950s. Its main idea was that parks and green areas were to be constructed primarily as social public places open for everyone. Parks, and open spaces, were designed to improve public health.[54] In a similar fashion, the role of sport in the decades after the war was increasingly recreational and accessible to a growing range of age and social groups. This was linked to an increase in leisure time of the working population, as well as a rise in the standard of living.[55]

In the municipal efforts to mobilise the masses an important part was played by the people's parks, which were under the jurisdiction of the Helsinki City Sports Department. The recreational role of sports areas became even more pronounced following the Great Annexation of 1946 after which the total area of Helsinki increased five-fold and the population grew by over 50 000 new inhabitants (with a total population 341,563 after the annexation). The Helsinki City Sports Department published a five-year plan in 1947 that stated its aim to make use of the newly acquired open green spaces to build more camping areas and cabins and to organise popular amusements in order to increase interest in outdoor activities.[56] The annexation brought Helsinki nine extra sports fields and five more swimming pools, but the city was faced with a large scale building project in renovating the newly acquired sports sites as well as building new ones.

The building of the suburbs began the process of decentralisation in Helsinki as city districts were no longer built as extensions to the old city layout but instead as separate areas 'in the middle of the forest'. This meant a new kind of challenge for the provision of sports facilities. The Helsinki City Sports Department plan of 1947 had as its special concern to increase the use of Helsinki's urban forests and special attention was paid to forest pathways and cross-country skiing tracks. The city also aimed at building

---

53 Aalto, 'Urheilu – viihteestä viihdeteollisuudeksi', 347–348.
54 L. Nilsson, 'The Stockholm Style: a model for the building of the city in parks, 1930s–1960s', in Clark (ed.), *The European City*, 142–143.
55 Aalto, *Helsingin kaupungin urheilu- ja ulkoiluhallinto*, 47.
56 Aalto, *Helsingin kaupungin urheilu- ja ulkoiluhallinto*, 44.

skating rinks in every part of the city. The next five-year plan by the Helsinki City Sports Department in 1958 had similar aims to the previous one. The emphasis was still on mass sport and the city's green spaces were seen as the most important element in helping to increase physical exercise in all age groups.[57]

Conclusion

Sports sites in Helsinki in the period from the late nineteenth century to the 1960s were highly contested space, and several key actors, private individuals, politically involved sports societies, teachers, in addition to the municipality, were involved in the shaping of sports activity and sports grounds. What is distinctive for this period is the growth of municipal involvement in the planning and building of sports sites, as well as its increasing interest in offering recreational activities for its inhabitants. This development culminated during the post-war era.

As we have seen, sports sites were heavily segregated space in Helsinki since sports grounds, even when owned by the city, were managed by and reserved for certain sports societies thus excluding other users. Another factor affecting space for sport in Helsinki was the gender aspect: outdoor sports grounds were mostly used by men thus largely excluding girls and women from using them. The Finnish school system further underlined the gender distinction, as before the school reform of the 1970s girls and boys were for the large part taught separately and followed different curricula; the curriculum in girls' schools contained significantly less sports education and less use of outdoor space than that in boys' schools. No less important for excluding females from sport and outdoor sports areas was the fact that competitive sport was largely regarded as more suited to males than to females. The class division in sport, that had been strong in the nineteenth century before sport became widely followed across all social classes, was replaced by another kind of division. From 1919 to 1993 the Finnish central organisation for sport, *Finnish Athletics Association* (SVUL) and the *Finnish Workers' Sports Federation* (TUL) separated the Finnish sports field into two separate and mutually conflicting divisions.

Staging the Olympics Games in Helsinki resulted in the building of sports sites fit for international competitive sports, but for many years afterwards decreased the municipal funds available for mass sport and exercise. Paradoxically it served not only to increase the division between elitist and mass sports but also stimulate wider public interest in sport. One legacy of the 1952 Summer Olympics in Helsinki was to boost the country's national image and to enhance the sense of national cohesion. In the case of Helsinki hosting the Olympics meant important changes in its landscape, making the new sports spaces, the Olympic village and other developments highly visible features of the modern city.

---

57 Aalto, *Helsingin kaupungin urheilu- ja ulkoiluhallinto,* 45.

Post-war Helsinki saw sport and green space as alleviating anti-social behaviour and preventing social unrest, along with the promotion of public health and social welfare. This social role of sport can also be seen in the five-year plans the Helsinki City Sports Division made in 1947 and 1958. They show how the aims of the city had moved towards supporting mass sport recreational activities. Similar aims were stated yet again in 1969 by Helsinki's Sports and Outdoor Administration: '....to create conditions where every resident, regardless of age and sex, has an opportunity of engaging in sport and outdoor recreation of various kinds and stages, according to his own needs, in forms which are acceptable both socially and with regard to health; it should also be mentally and physically refreshing'.[58] As Suvi Talja shows in the next Chapter on green space and sport in Helsinki since the 1960s, the municipal aim in providing sport had found its main emphasis, which was to promote the idea of 'sport for all'.

---

58 Aalto, *Helsingin kaupungin urheilu- ja ulkoiluhallinto*, 85.

SUVI TALJA[1]

3 Sport for All? – Development of Sports Sites and Green Space in Helsinki Since the 1960s

Increased urbanisation in Finnish society, along with greater leisure time in the 1960s, resulted in growing recognition of the recreational needs of a wide range of people, not only those involved with competitive sports. There was pressure to distinguish between different types of physical activities and sports, and thus acknowledge the diversity in sporting habits of people. Sports culture was differentiated between competitive sports at the top level, amateur competitive sports, sports for health and sports education. The Finnish government launched a national policy to support sport for the masses, or 'sport for all', as the term has been used in a wider European movement.[2]

The Finnish economy experienced steady growth from the 1950s until the early 1990s, and public funding for sports facilities increased.[3] Municipalities, with the help of growing subsidies from the government, invested in sports sites, such as recreation routes, ballgame grounds and swimming halls. High-standard venues for top-level competitive sports were likewise funded from public sources. As Katri Lento discussed in Chapter Two the sports landscape[4] of Helsinki already experienced significant changes before and after the Second World with the construction of sports facilities for the Olympic Games in 1952. These high quality facilities were built in

---

1 I am grateful for funding for this work to the European Studies Network project *Green Space, Sport and the City*.
2 S. Hentilä, 'Urheilupolitiikasta liikuntapolitiikkaan', in T. Pyykkönen (ed.), *Suomi uskoi urheiluun: Suomen urheilun ja liikunnan historia*. Helsinki: VAPK-kustannus 1992, 354–356. When addressing the recreational and non-professional aspect in sports, the term 'liikunta' ('movement') started to be preferred over 'urheilu' ('sport'). The idea of 'Sport for All' was adopted in 1966 by the Council of Europe. See *Sport for All: Five Countries Report*. Strasbourg: Council of Europe, Council for Cultural Co-operation 1970. In Finnish, 'sport for all' was often translated as 'kuntoliikunnan edistämistyö' or 'kuntoliikuntaa kaikille'. See for example J. Juppi, 'Eurooppalaista "Sport for All" toimintaa', *Stadion* 5–6/1973.
3 J. Juppi and K. Ilmanen, 'Suunnittelu-uskosta verkottumiseen, aluepolitiikasta yksityistämiseen, perustoiminnasta projekteihin', *Liikunta ja Tiede* 6/1999.
4 The concept of 'sports landscape' or 'scenery of sports' can be interpreted as a type of cultural landscape including people and buildings in a sports context. See J. Bale, *Landscapes of Modern Sport*. Leicester: Leicester University Press 1994, 9.

the centre of Helsinki, while the suburbs annexed to the city in 1946 had none or very few built sporting facilities.

This chapter focuses on the development of new sports sites after the 1960s, and changes in sports culture in Helsinki. It looks in turn at the emergence of the concept of 'sport for all' and then in more detail at the development of sports sites in Helsinki. Questions addressed include: what type of spatial structure was promoted by city officials and planners? What has been the change in indoor and outdoor sports, and how has the commercialisation of sports culture manifested itself in Helsinki? Finally, the chapter examines the challenges and problems in developing sports sites under the idea of 'sport for all', not least in the recent discussion of the relationship between sport and environment.

Welfare state and the idea of 'sport for all'

Finnish society was rapidly urbanising in the 1960s due to massive immigration from the countryside. Jobs in trade and services replaced those in agricultural sector, and by the 1970s only 20 per cent of the working population had a manual job. Meanwhile, the amount of leisure time was increasing, especially after 1965, when the 40-hour work week was introduced. These changes encouraged greater consciousness of the importance of sport and its societal function. Various leading figures promoted sports for health and urged a better balance between public funding for top-level competitive sports and sports for a wider public. Among them, the Finnish sports sociologist Kalevi Heinilä and President Kekkonen, an active sportsman himself, promoted the importance of exercising.[5] It was also recognised that people increasingly wanted to exercise individually or informally rather than as a member of a sports organisation. According to a national study at the beginning of the 1970s, 61 per cent of the Finnish people over 15 years old exercised, but only 17 per cent were active in sports organisations.[6] The old ideology of sports policy was becoming less relevant. In many of the constructed sporting sites, the main idea and design in their establishment was to support the needs of competitive sports.[7] The Finnish journal, *Stadion*, stated in 1968: 'The pro-competitive hegemony has produced more seating for spectator sports than space for actual physical activity practised by the wider public'.[8] Finland was, and to a some extent still is, a good example of a country where celebrating famous sportsmen,

---

5   J. Juppi and J. Aunesluoma, 'Valtiovalta ja liikunta 1920–1994', in J. Aunesluoma (ed.), *Liikuntaa kaikelle kansalle: Valtion, läänien ja kuntien liikuntahallinto* 1919–1994. Helsinki: Liikuntatieteellinen Seura ry 1995, 73–75.
6   Hentilä, 'Urheilupolitiikasta liikuntapolitiikkaan', 361.
7   Niemen toimikunta, *Kuntoliikunnan suunnittelu: Sosiaaliset edellytykset ja kuntoliikuntapaikat. Niemen toimikunnan osamietintö*. Komiteanmietintö B:1968: 3. Helsinki 1968, 8; Juppi and Aunesluoma, 'Valtiovalta ja liikunta', 73–75.
8   P. Kiviaho, 'Virkistys- ja kuntoliikunnan kunnallisesta suunnittelusta', *Stadion* 1/1968, 9–12. *Stadion* changed its name to *Liikunta ja Tiede* in 1981.

and more increasingly also sportswomen, has been an important dimension of national self-definition.[9]

Along with the municipalities, the Finnish government became more involved in planning, administering and funding sports for health in the 1960s. Developing public sports policy was regarded as an important part of the welfare state.[10] One of the government's actions was to appoint a committee (Niemen toimikunta) in 1966 to examine how 'sport for health' and 'sport for all' should be promoted in Finland.[11] Also, the space for physical exercise for children was given added attention by the national agency responsible for education. In 1979 normative guidelines for school yards and spaces for recreation were established. Guidelines assigned minimum standards for the type and size of playing fields and recreation routes and their distance from school.[12]

At a national level, the process of developing 'sport for all' culminated in the Sports Act of 1980. This changed the municipal sports service from a discretionary to a statutory policy in all municipalities in Finland. The main guideline was that: '... creating the general prerequisites for sports and physical activity is the task of the government and municipalities, whereas the actual organising of the sports activities is the task of sports organisations'.[13] With the Sports Act, the government encouraged municipalities to design more comprehensive and long-term sports plans. These plans were especially required when municipalities applied for state subsidies for big sports site developments.[14] Also, the new Act recognised special groups, such as disabled and elderly as important subjects of public sports policies.[15]

The network of sports sites

In the late 1950s, the Sports Department of the city of Helsinki and a special Leisure Committee, set up by the city government in 1956, formulated long-

---

9 M. Tervo, *Geographies in the Making: Reflections on Sports, the Media, and National Identity in Finland*. Oulu: Geographical Society of Northern Finland 2003, 1–3; H. Meinander, 'Prologue: Nordic History, Society and Sport', in H. Meinander and J. A. Mangan (eds), *The Nordic World: Sport in Society*. London: Cass 1998, 5.
10 Juppi and Auneslouma, 'Valtiovalta ja liikunta', 74.
11 Juppi and Auneslouma, 'Valtiovalta ja liikunta', 75; P. Oja, 'Urheilun suunnittelu nykyaikaistuu – kuntoliikunnalle suunnitteluperusteet 1970-lukua varten', *Stadion* 1/1968.
12 *Peruskoulun tontin suunnitteluohje*. Yleiskirje 3045. Kouluhallitus 1979, as cited in *Koulupihojen liikuntaolosuhteet: Valtakunnallinen tutkimus 2003*. Nuori Suomi ry 2004, 12. The national standards for school yards were abolished in 1991.
13 Although providing sports facilities and services was not a statutory policy before 1980, there was a sports board or similar in 96 per cent of the municipalities. The new law basically improved the functioning possibilities of the sports boards and strengthened their administrative position. K. Ilmanen, 'Kunnallinen liikuntatoimi 1919–1994', in Auneslouma (ed.), *Liikuntaa kaikelle kansalle;* J. Juppi, 'Lakisääteistä liikkumista vuodesta 1980', *Liikunta ja Tiede* 2/1997.
14 P. Salminen, *Urheilualuetutkimus: Liikuntapaikkojen suunnitteluohjeet*. Opetusministeriö, urheilu- ja nuoriso-osasto 1981, 90–91.
15 K. Ilmanen, *Ensimmäisenä liikkeellä: Helsingin kaupungin liikuntatoimi 1919–1994*, Helsinki: Helsingin kaupungin liikuntavirasto 1994, 163–164, 202.

term proposals for leisure and sports policy in Helsinki. The suburbs annexed to the city in 1946, accommodating the new housing areas, were given official priority in terms of accessible sports and recreation sites.[16] The emphasis was on turning green areas into spaces for different sports activities for the masses. Recreation routes and, in wintertime, skiing tracks would be built and lighting provided.[17] Constructed sporting facilities would be organised as a spatially balanced, hierarchical network. The network would consist of three main types of sports sites: neighbourhood playing fields adjacent to apartment blocks, suburban playing fields, and sports parks. The densest network would consist of neighbourhood playing grounds for young children. Bigger suburban playing grounds were to serve school children in particular, and so be adjacent to schools. In addition, spaces for special sports, such as long jump sites, would be built there. Such sites were to serve other users outside school hours. At the top of the hierarchy were the sports parks, accommodating various facilities for different sports, and serving larger areas with a population of at least 50,000. Suburban playing fields and sports parks were to be connected to wider green areas by recreation paths.[18] In public discussions, it was stressed that even the planned big sports parks would serve 'all the people', not just those involved in top-level competitive sports.[19]

The idea of a hierarchy of sports parks and other exercising sites seems to have followed general planning ideas in the capital area in the 1960s,[20] and was included in the Proposed Helsinki Master Plan in 1960. The plan

---

16 However, already in 1947, the Sports Department stated that the municipality should engage with promoting sporting activities for large masses, and professional sports could be left for sports associations and federations. See T. Aalto, *Helsingin kaupungin urheilu- ja ulkoiluhallinto 1919–1969*. Helsingin kaupungin julkaisuja 22. Helsinki 1969, 42. The new areas annexed to the city in 1946 had very few built sports sites. A survey of the sports sites in the new areas concluded that the best facilities were in Kulosaari, a villa community for the upper class. Kulosaari had tennis courts, a sports ground, a golfing area, a beach with pavilions, and an ice-skating rink with light. Typically, the annexed communities had only one athletics field or sports ground, and some had a beach. See Ilmanen, *Ensimmmäisenä liikkeellä*, 90.
17 Aalto, *Helsingin kaupungin urheilu- ja ulkoiluhallinto*, 45.
18 Helsingin Liikuntaviraston arkisto (The City of Helsinki Sports Department Archives), Urheilu- ja ulkoilulautakunta, Me:17, Vapaa-ajanviettokomitean mietintö 21.9.1962, 25–26. The idea of a systematic network of urban parks connected by green corridors had originally appeared in the mid-nineteenth century and advocated by the planner of the Central Park of NewYork, architect Frederick Law Olmstead. In Europe, open and green spaces were an important part of the work of the British architect and town planner Patrick Abercrombie, who had a great influence on the design of London after the Second World War. In Finland, in the 1910s, the Finnish architect Bertel Jung argued for the importance of sports facilities in urban areas, and the idea of placing sports sites within parks and nature. Another Finnish architect, Otto-I. Meurmann, wrote in a similar way about sport and parks in 1947 in his influential book on planning, *Asemakaavaoppi*.
19 'Urheilupuistoajatus edistää kokonaisliikuntaa, pitkäntähtäimen ohjelma helpottaa taloutta', *Helsingin Sanomat* 16.2.1967.
20 In the 1960s, the idea of a hierarchical central network was promoted to guide all urban development, for example building town centres, 'New Towns', with 20,000–50,000 inhabitants. See H. Schulman, *Alueelliset todellisuudet ja visiot: Helsingin kehitys ja kehittäminen 1900-luvulla*. Espoo, Teknillinen korkeakoulu 1990, 120.

proposed eight areas for sports parks varying from 15 to 60 hectares, most of which were entirely new and were to be located in the newly annexed areas of the city. Three of the new planned sports parks were especially important: Tali (serving the west of the city), Pirkkola (serving the north) and Myllypuro (serving the east).[21] The Master Plan of 1960 did not have a big effect on city planning in Helsinki, but the main idea of saving the green corridors and establishing connecting recreation routes was incorporated into subsequent plans with more significance, notably the Master Plan of 1972 and the Regional Plan of 1975.[22] Though urban densification since the 1970s has diminished the amount of green space in Helsinki, the earlier planning ideas have survived into the twenty-first century: thus the Master Plan of 2002 emphasised preserving the existing green corridors of Helsinki as a network connecting different housing, recreation and built sports park areas.[23]

The changing character of sports sites

'The construction work for the swimming and ballgame hall required such a large pit for its foundation that it seemed the whole forest had been turned upside down.' This recollection from a resident who lived near the new Pirkkola sports park in the late 1960s illustrates the new scale of sports building in Helsinki.[24] Creating spaces for active sports often diminished the amount of green space rather than the other way around. However, one could also claim that there was more than sufficiently green space in Helsinki to accommodate major sports complex developments.

In the 1950s, the most common built sports site in Finland was the athletics field. In the following decade, however, improved living standards and greater concern for personal health significantly boosted public demand for sports and recreation possibilities. In 1960 the National Sports Board and the Regional Sports Boards stated the need to shift funding from athletics fields to a wider range of provision, including both outdoor and indoor sporting facilities.[25]

21 T. Herranen, 'Kaupunkisuunnittelu ja asuminen', in O. Turpeinen, T. Herranen and K. Hoffman, *Helsingin historia vuodesta 1945:1*. Helsinki: Helsingin kaupunki 1997, 211; *Helsingin yleiskaavaehdotus: Laadittu asemakaavaosastolla 1953–1960*. Helsingin kaupungin julkaisuja N:o 9. Helsinki 1960, 66. In addition to Tali, Pirkkola and Myllypuro, the plan proposed two smaller parks for the Malmi area, and one to Herttoniemi. Small sports parks were also proposed for Lauttasaari, Oulunkylä and Laajasalo. In the central Helsinki (kantakaupunki) there were already two sports parks around the Stadium area and Käpylä.
22 H. Schulman, 'Helsingin suunnittelu ja rakentuminen', in H. Schulman, P. Pulma and S. Aalto, *Helsingin historia vuodesta 1945:2*. Helsinki: Helsingin kaupunki 2000, 57, 66.
23 Herranen, 'Kaupunkisuunnittelu ja asuminen', 212–213; Yleissuunnitteluosasto, *Helsingin yleiskaava 2002, ehdotus, selostus*. Helsingin kaupunkisuunnitteluviraston julkaisuja 2002:17, 157–163.
24 Helsingin kaupunginarkisto (Helsinki City Archives), Elämää lähiössä – written memories, collected by *Helsingin Sanomat* from people who lived in the new suburbs in the 1960s and 1970s, Pohjois-Haaga.
25 H. Klemola, 'Läänien liikunta ja urheilu 1946–1994', in Aunesluoma (ed.), *Liikuntaa kaikelle kansalle*, 164.

In Helsinki, the number of ballgame grounds grew strongly from one hundred in the early 1960s to several hundred in subsequent decades.[26] Because of the Finnish climate, grass fields were not easy to maintain and thus grounds were mostly sand-based. In the wintertime, they were frozen to serve as skating rinks.[27] Recreation in natural green space was also made more accessible. Within the city borders, there were 5,740 hectares of green space, consisting of mainly forests. Roughly one third were cemeteries and agricultural fields. The total amount of green space comprised 34 per cent of the land area in Helsinki.[28] Several city departments were involved in the management of these different types or green areas.[29] By the end of the 1950s it was recognised that the old 'people's parks' (kansanpuistot), such as Seurasaari and Kivinokka (see Chapter Two by Katri Lento), were suffering from overcrowding. To improve provision for outdoor recreation, the city of Helsinki bought extensive green areas from the surrounding communities.[30] In the 1970s new recreation and jogging routes were built near housing areas.

Helsinki residents could also swim outside in a natural environment, at least during the summertime. After the annexations of 1946 and 1966, Helsinki had a coastline of about 98 kilometres, and a large archipelago, in part open to the public. However, in the 1960s, most of the coasts were reserved for industries or harbours, or were suffering from pollution from sewage. This was especially a problem for downtown areas of Helsinki,[31] which were often the most popular places for summer time leisure.[32] The two large outdoor swimming complexes built for the Olympics of 1952 were also actively used in summertime.[33] The Leisure Committee argued for new open outdoor lidos, along with some indoor pools, in its report of 1962. Increasingly, from the mid-1960s, the city authorities were concerned to provide sufficient spaces for year-round swimming in indoor pools. As well as answering the problems of a short summer, sea pollution and the

---

26 Ilmanen, *Ensimmäisenä liikkeellä*, 111. In 2005, the different ballgame grounds in Helsinki consisted of 364 sand fields, 81 grass fields and 7 sand-based grass fields, the administration of which was divided between the Sports Department, the Public Works Department and the Education Department, the first two being in charge of the majority of these sports grounds. These figures are withdrawn from the national databank for sports sites, *Suomalaisen liikunnan tietopankki: Vertti-raportti*. www.sport.jyu.fi.
27 *Kertomus Helsingin kaupungin kunnallishallinnosta 86: 1973*. Helsinki: Helsingin kaupungin tilastotoimisto, 1976, 18–19.
28 *Helsingin yleiskaavaehdotus*, 65–67.
29 L. Lehto, *Elämää helsinkiläismetsissä: Helsingin kaupunginmetsänhoidon syntyhistoriaa*, Helsinki: Helsingin kaupunginmuseo 1990; S. Simonen, *Helsingin kaupungin rakennusvirasto 100 vuotta*. Helsingin kaupungin julkaisuja 31. Helsinki 1978, 133–135.
30 Ilmanen, *Ensimmäisenä liikkeellä, 97–100*.
31 *Helsingin yleiskaavaehdotus*, 67. According to an article in newspaper published at the time by the Communist Party of Finland and the Finnish People's Democratic League, *Kansan Uutiset*, the three most polluted beach areas were close to downtown Helsinki: beaches at Mustikkamaa, Kulosaari and Hietaranta. Also, one of the swimming areas in western downtown, Humallahti, had been restricted from swimming in 1958. 'Helsingin uimavedet – likaisia mutta pikkuhiljaa paranemassa', *Kansan Uutiset* 28.6.1974.
32 'Helsingin kesä: Rantaa riittää, mutta ei uimarantaa', *Kansan Uutiset* 26.5.1977.
33 P. Salminen, 'Liikuntalaitosten suunnittelu', *Arkkitehti* 4/1977, 14–18.

growth of inland residential areas far away from suitable beaches, indoor pools were also regarded important for educational purposes. The new sports parks were first seen as the most suitable places to accommodate new indoor pools.[34] As the article in *Helsingin Sanomat* in 1967 stated: '... not until there are swimming halls in Pirkkola, Myllypuro, Tali and Malmi-Oulunkylä sports parks, can one be satisfied with the year-round recreational and educational swimming possibilities in the capital area'.[35] In 1973, the city established a committee to consider comprehensive programme for the building of indoor swimming halls throughout the city. Altogether, between 1960 and 1980, there were seven new indoor swimming halls established, with various funding patterns from public and private sources.[36] As we will see below, despite plans for comprehensive swimming pool network in Helsinki, the provision of indoor pools across the city was slow to occur and spatially unbalanced.

The building of different indoor facilities accelerated in the 1980s, and this development continued into the 1990s.[37] According to Kalervo Ilmanen, there were 187 indoor facilities in 1980 and 350 in 1990. But outdoor sports areas did not lose their popularity.[38] Particularly important here was the growth of more informal individual sports, such as jogging and walking. According to a 1986 Helsinki city statistics survey, 80 per cent of Helsinki residents exercised informally outdoors.[39] That year, the city had 300 kilometres of recreation routes, of which 70 per cent were in urban forests or in other recreational areas, though the fact that only two thirds of them were illuminated hindered their use during dark autumn and winter evenings.[40]

Pedestrian ways and streets were also increasingly used for outdoor exercising, and sporting events such as marathons became very popular. For instance, the number of registered participants in the Helsinki City Marathon, established at the beginning of the 1980s, grew from 3,000–4,000 participants during the early years to more than 6,000 in 2007; here commercial sponsorship was increasingly important.[41] Another mass event,

34 Vapaa-ajanviettokomitean mietintö 21.9.1962, 31–36; A. Numminen, 'Kunnallisen urheilu- ja liikuntatoiminnan piirteitä Helsingissä', *Stadion 1/1968,*
35 'Urheilupuistoajatus edistää kokonaisliikuntaa, pitkäntähtäimen ohjelma helpottaa taloutta', *Helsingin Sanomat* 16.2.1967.
36 Three of the new indoor pools were built adjacent to schools and one adjacent to a rehabilitation centre. Helsingin kaupunginarkisto (Helsinki City Archives), Helsingin kaupunginvaltuuston asiakirjat, Helsingin kaupunginhallitus, mietintö nr. 4 – 1974, Liite: Helsingin kaupunginhallituksen asettaman uimalakomitean mietintö 30.5.1974, 5–15.
37 *Helsinki tilastoina 1800-luvulta nykypäivään.* Helsinki: Helsingin kaupungin tietokeskuksen tilastoja 15/2000, 138; L. Kähkönen, *Kulttuuri- ja vapaa-aikatoimen 1990-luku.* Helsinki: Helsingin kaupungin tietokeskuksen julkaisuja 2004, 7.
38 Ilmanen, *Ensimmäisenä liikkeellä,* 162.
39 *Kulttuuri ja vapaa-aika Helsingissä, Tilastoja 1989 E:1.* Helsinki: Helsingin kaupungin tietokeskus 1989, 121.
40 *Ulkoilureittien toteuttamisohjelma: Pääulkoilureitistön toteuttamisohjelma perusteluineen vuosille 1987–1993 sekä ehdotukset jatkotoimenpiteiksi.* Helsinki: Helsingin kaupunki 1987, 4, 17.
41 Website for Helsinki City Marathon, Historiikki. <http://www.helsinkicitymarathon.fi/fi/historiikki.html>. Visited 20.11.2008.

albeit less competitive was launched in the early 1980s. The womens' ten kilometre race (Naisten Kymppi) began as a social gathering, and not as a serious-minded competition. Starting with fewer than 400 female participants in 1984, by 1990 it had turned into a huge mass event of 30,000 runners, and still remains an important event in Helsinki.[42]

In general, public expenditure on sport grew steadily from the mid-1960s until the economic crisis of the 1990s. The number of sports sites per inhabitants in Helsinki roughly tripled during this period.[43] From the 1980s there was also an important growth in commercial sports providers.

Commercialisation of sports versus public provision

The booming economy of Finland in the 1980s spurred private sports developments. Exemplifying the new commercialisation of sports, trendy gyms started to advertise their facilities in Helsinki newspapers.[44] The majority of the new trend sports facilities of the 1980s, such as tennis- and squash-halls, aerobics studios and down-hill skiing centres were provided by private entrepreneurs. There was also a boom in building private golf courses in Finland (see Chapter Nine) in the 1980s. Between 1982 and 1991 six new full-size golf courses were established in the metropolitan area, though none inside the city limits.

No studies currently exist of the development of commercial sporting venues in Helsinki, and due to their private ownership and high turnover it is hard to identify long trends. Teemu Aalto listed 15 'private indoor halls' for gymnastics or other sports in 1967 in the official history of the Sports Department in Helsinki, but it is unclear whether these were all commercial venues.[45] Another survey on leisure patterns of Helsinki residents in 1989 reported 58 commercial sports enterprises offering instructed activities, such as different types of aerobics, dances, weight lifting and tennis courses. Commercial venues were most likely on the increase from the 1980s. Doing aerobics or going to the gym was seen as something 'urban', appealing to those who saw their image as 'young, beautiful and successful'. An article in the daily newspaper *Uusi Suomi* stated in 1987: 'Cross-country skiing, traditional gymnastics and team ball games are sports for ancestors, now everyone up to date must have a club card to a fitness centre…'.[46] Even if this was a provocative exaggeration, sports culture was undoubtedly diversifying and becoming part of consumer culture in the 1980s.

Despite the growth of commercial and private provision of individual sports, the Sports Department was also responding to the new trends. For

---

42 R. Väinämö, 'Kuntoklubeista naisten vaikutuskanavia', *Helsingin Sanomat* 25.1.1991.
43 Ilmanen, *Ensimmäisenä liikkeellä*, 162. Sports sites include in this calculation for instances skiing tracks and recreation routes, the importance of which would be better evaluated if examined by their length.
44 *Kulttuuri ja vapaa-aika Helsingissä*, 123; S. Aalto, 'Urheilu – viihteestä viihdeteollisuudeksi' in Schulman, Pulma and Aalto, *Helsingin historia vuodesta 1945:2*, 407.
45 Aalto, *Helsingin kaupungin urheilu- ja ulkoiluhallinto*, 75.
46 *Kulttuuri ja vapaa-aika Helsingissä;* S. Tikanoja, 'Kohteena on keho', *Uusi Suomi* 26.4.1987.

example, in the mid-1980s the Department started to arrange aerobics and gymnastics sessions in those parks with large grass lawns in the summertime. The first parks to accommodate outdoor aerobics were downtown, in Kaisaniemi and Hesperia, but currently also several parks in the suburbs arrange these activities.[47] Outdoor aerobics was popular among women, and was seen to combine the aesthetic enjoyment of the natural surroundings and physical activity.

There was also greater diversification in case of sports associations. In Helsinki, the number of sports clubs (receiving subsidies from the city) increased from 135 in 1960 to 420 in 1990.[48] The main reason for this increase was the growth of club specialisation as earlier sports fragmented into numerous new activities. Commercial sponsorship money for club activities also became more common.[49] This trend was especially seen in the top-level competitive sports, but also occurred in clubs playing in lower divisions and clubs for adolescents and children.

The economic recession in Finland at the beginning of the 1990s had a serious impact on public sports policy. State subsidies as well as the municipal funds for sport were cut. In the mid-1980s there were over 400 municipal sports boards (practically one in every municipality), but in 1994 only about 20 boards were left – the others were usually merged with other boards, such as boards dealing with culture or adolescents.[50] In Helsinki, funding for the Sports Department was reduced at the beginning of the 1990s, which led to a decrease of personnel and investment in new facilities, though funding was increased again in the mid-1990s. At that time, a major expenditure for the city (with some funds from the government) proved to be the repair and restoration of the sports facilities built for the Olympics in 1952. Thus, the era of building new facilities had turned, at best, into an era of refurbishment.[51] Funding for parks under the jurisdiction of the Public Works Department experienced a steady decline from 1991 to 1997, resulting in the dilapidation of those parks needing intensive maintenance.[52] However, compared to some smaller cities or rural municipalities in Finland, the city of Helsinki was able to maintain its public sports facilities and parks relatively well. Furthermore, although commercial sports venues have been on the increase, the majority of built sports sites are currently run by the Sports Department and practically all the public parks with woodlands, meadows or grass lawns are managed by the Public Works Department.

---

47 M. Nyström. Sport planner, City of Helsinki Sports Department. Email communication. 1st November 2007.
48 Ilmanen, *Ensimmäisenä liikkeellä,* 178, 121.
49 Aalto, 'Urheilu – viihteestä viihdeteollisuudeksi', 478
50 K. Wuolio, 'Sivistystoimentarkastaja Antti Lehtinen: Kunnissa enemmän edunvalvojia kuin toiminnan kehittäjiä', *Liikunta ja Tiede* 2/1999; K. Sjöholm, 'Raunioitetaanko liikuntapaikat?', *Helsingin Sanomat* 11.7.1994.
51 Ilmanen, *Kunnallinen liikuntatoimi,* 314–315.
52 Helsingin Rakennusviraston arkisto (The Archives of the Public Works Department, City of Helsinki), The annual reports of the Public Works Department 1991–1998; H. Värri, Head of Office, Park Division of the Public Works Department, personal communication 16.5.2006.

Sport for all? – Challenges in the sports site development

So far we have been analysing the general trends in sports provision in late twentieth-century Helsinki, but as we will see there were also important variations in provision both in terms of spatial area, types of facility, and the gender of participants. The first of the new planned sports parks was the Pirkkola sports park, mainly constructed by public funding between 1965 and 1974 in the middle of the Central Park of Helsinki. First, there were built facilities such as an artificial skating rink and a sand football field. By 1969, provision was made for a grass football field, an athletic field, a hall for ballgames and swimming, and twelve kilometres of lit recreation routes around the built sporting facilities. These routes were maintained as ski tracks in winter.[53] Pirkkola was located in the northern area of Helsinki and was accessible from a large number of nearby middle-class suburbs. At that time, the northern part of the city was seen to lack proper sports and leisure facilities. Already in 1958, when an architectural competition was launched for the Pirkkola sports park, the surrounding suburbs were active advocates for building leisure and sports facilities in the area.[54]

The next sports parks in the planning list were Myllypuro, Tali and Oulunkylä, but it took much longer to establish them. Completing the Pirkkola sports park and building an indoor ice stadium in the centre of Helsinki in 1966 left very little tax money for other immediate sports site developments.[55] The Myllypuro sports park was planned to serve the eastern part of the city, where there was the greatest need for sports facilities at the time. That part of the city accommodated a number of new suburbs built in the 1960s and 1970s. These 'forest suburbs' were characterised by high-rise apartment blocks, mainly for lower income groups, with relatively poor public transportation connections. This was not a unique situation in the Nordic countries; at that time the city of Stockholm saw a very similar pattern.[56] The provision of services and amenities, such as sports facilities, often arrived long after the housing development.[57]

---

53 Ilmanen, *Ensimmäisenä liikkeellä*, 140–141; Aalto, *Helsingin kaupungin urheilu- ja ulkoiluhallinto*, 56. There was already a pool with natural boundaries in the park area.
54 'Pirkkolan urheilupuiston aatekilpailu', *Arkkitehti*, 8/1959, Kilpailuliite no. 1; 'Peräänantamattomuus toi voiton Pirkkolan urheilupuistolle', *Liitosalueen sanomat* 25.4.1958; Helsingin kaupunginarkisto (Helsinki City Archives), Kaupunginvaltuuston keskustelupöytäkirja 26.4.1961, 313 §, Esityslistan asia n:o 20.
55 Ilmanen, *Ensimmäisenä liikkeellä*, 142–144. Building of the indoor ice-skating rink cost the city as much as the entire Pirkkola sports park. It is notable, that Helsinki received relatively little funding for the ice stadium from the Finnish government compared, for example, to Tampere, which also built a new indoor ice hockey rink at that time.
56 L. Nilsson 'The Stockholm Style: a model for the building of the city in parks, 1930s–1960s', in P. Clark (ed.), *The European City and Green Space: London, Stockholm, Helsinki and St Petersburg, 1850–2000*. Aldershot: Ashgate 2006, 155.
57 I. Roivainen, 'Metsäiseen kalliomaastoon 10 000 asukkaan kaupunki: luonto helsinkiläisten lähiökuvauksissa', in S. Laakkonen, S. Laurila, P. Kansanen and H. Schulman (eds), *Näkökulmia Helsingin ympäristöhistoriaan: Kaupungin ja ympäristön muutos 1800- ja 1900 -luvuilla*. Helsinki: Helsingin kaupungin tietokeskus 2001, 144–150. See also, 'Urheilukenttien puute harjoittelun esteenä Kontulan seudulla', *Helsinki–lehti* 16.8.1968.

In the case of Myllypuro, the first plan for the whole sports park was already finished in the early 1960s. The proposals included an indoor swimming pool as a part of a three-storey sports hall complex. Construction was postponed, but in 1974 a new town plan for the sports park was completed and again a swimming hall was prioritised. At that time Kontula, a suburb east of Myllypuro, was the only place in the whole of eastern Helsinki that had a swimming hall.[58] Once more, however, the building of the Myllypuro swimming hall was heavily delayed. Lack of funds at the Sports Department played a part but there was another reason for the delay: other sports sites in the city succeeded in gaining priority. For instance, a large share of public sports funds were allocated to an artificial ice-skating rink in Oulunkylä sports park (the largest of its kind in Finland that time); an area again in Northern Helsinki.[59] The swimming hall in Myllypuro had to wait until 1993, when the Sports Department built the facilities adjacent to a nearby shopping and office center at Itäkeskus, south of the sports park.[60]

Myllypuro seems to have been one of the suburbs that lacked large scale sports facilities for a long time, despite plans from the 1960s. At the end of the 1960s, the only built sports site in Myllypuro was a 0.3 hectare sandy ballgame ground adjacent to a school. In winter, the ground was partly frozed to serve as a skating rink.[61] There was also an unofficial site for football; a local football club, *Myllärit*, had voluntarily cleared a vacant lot suitable for playing. Additionally, some of the open spaces such as meadows and old agricultural fields offered unofficial practising sites for javelin, long-jumping and volleyball.[62] At the same time, the surrounding forest area was used informally for many other activities. In recorded memories of suburban living, collected by the main daily newspaper *Helsingin Sanomat* in 1995, the surrounding forests were often mentioned as important space for individual sports such as jogging and skiing. The following quotes are from the material collected from Kontula suburb. One woman recalled: '…we used to ski on the skiing tracks made in the nearby forest…', and according to another woman: 'In the wintertime, one could go skiing right from one's front door. In summer there was a lot of forest to use for recreation.' The same women continued: 'for youngsters, there were weights [for body building] in the commons area, but those days that was only a hobby for the boys… playing football on the grass lawns was prohibited … boys tended to play it until they were stopped. But there was also a proper playing ground nearby.' The beaches in eastern Helsinki were clean at that time, so they also offered recreation spaces for those living nearby or willing to take time going there.

58 Myllypuron urheilupuisto, asemakaava ja asemakaavan selostus. Helsingin kaupungin kaupunkisuunnitteluvirasto, Itäkeskus-projekti, Helsinki 1974, 31; 'Urheilupuistoaloite: Uimahalli kiireellisin kohde Myllypurossa', *Kansan Uutiset* 25.5.1974.
59 'Myllypuron uimahalli viivästyy', *Helsinki-lehti* 3.3.1977; 'Maauimalat Munkkiniemeen, Mustikkamaalle, MaTaPuPuun', *Helsinki-lehti* 22.8.1974.
60 Ilmanen, *Ensimmäisenä liikkeellä*, 145.
61 Helsingin Liikuntaviraston arkisto (The City of Helsinki Sports Department Archives), Liv, ulkoilu- ja urheilupalvelut 1969 Uj:20.
62 A. Valkonen, *Myllypuron kartanon takamailta monikulttuuriseksi kaupunginosaksi*. Helsinki: Myllypuro-seura 2005, 123–125.

As one woman recalls: 'In summertime, on weekends and sometimes weeknights, we went to the sand beach of Kallahti to wade, swim, read, have a coffee and admire the beautiful nature there.' Also, a woman who was a child in the late 1970s associated both forest and some built spaces to leisure time activities: 'There was a recreation route of two kilometres in the forest of Kurkimäki we used for walking and jogging with mum and dad.'...'With dad we went swimming twice a week, either to the Kontula swimming hall or the Pirkkola swimming hall... For a couple of years I took ballet courses, organised in a basement apartment ... and figure skating in the Myllypuro semi-indoor skating-rink'.[63]

Thus the lack of built sports sites in some poorer parts of the city was partially offset by the abundant green space, or other vacant spaces offering possibilities for informal activities. Adolescents, however, were regarded as suffering from the lack of organised activities. The high number of adolescents of a similar age in the new suburbs resulted sometimes in social unrest and vandalism, even gang fights. As other studies of Helsinki suburbs and that by Katri Lento in Chapter Two point out, an important reason to offer sports sites for young people, and especially young men, was to alleviate social problems.[64] The problem of offering sport for all was not only a spatial issue, however.

Competitive versus informal sports

Despite the increasing resources for the sports sector in municipalities, the implementation of the Sports Act of 1980 was criticised for neglecting those residents not active in sports organisations. For instance, in 1986 the journal *Liikunta & Tiede* (formerly *Stadion*) raised the question of whether contemporary public sports policy was able to meet the needs of an average person wanting to do sports individually, or whether the authorities dealing with sports in municipalities and government were still too favourable towards competitive and top-level sports.[65] One of the reasons for such a bias may have been the way that sports organisations had been heavily involved in the formulation of the Sports Act.[66] Furthermore, it was often representatives of the local sports organisations who were elected to the municipal sports boards. In practice, however, Finnish people increasingly

---

63 Helsingin kaupunginarkisto (Helsinki City Archives), Elämää lähiössä -materiaali, Kontula. See also K. Saarikangas, 'Nuoruus lähiössä', in K. Häggman (ed.), *Täältä tulee nuoriso! 1950–79*. Helsinki: WSOY 2006, 39–63. The last quote from a girl living in Myllypuro is apparently from the late 1970s. The Myllypuro sports park received its first built sports facility in 1976 when a non-profit company Nuorisojääkenttä Oy built a roofed skating rink there.
64 For example Aalto, 'Urheilu –viihteestä viihdeteollisuudeksi', 467–468.
65 T. Haukilahti, 'Miten liikuntakulttuuria kehitetään julkisessa hallinnossa?', *Liikunta ja Tiede* 5/1986.
66 P. Vuolle, 'Osallistuminen organisoituun liikuntaan', in P. Vuolle, R. Telama and L. Laakso (eds), *Näin suomalaiset liikkuvat*. Liikunnan ja kansanterveyden julkaisuja 50, Helsinki 1986, 144.

did sports alone or in groups, outside sports organisations.[67] In 1988, a national survey revealed that 73 per cent of respondents did sports individually. The next most common ways to do sports were through a workplace (9 per cent), a sports organisation (8 per cent) or a private company offering sports services (10 per cent).[68]

Compared to other cities in Finland, such as Tampere and Oulu, Helsinki had fewer sports sites per person. Kalervo Ilmanen has argued that this was because Helsinki had a denser and faster growing population and also because the city had larger, high quality sports facilities.[69] The design and purpose of newly built sports sites in the city were frequently criticised for emphasising too much the competitive aspect in sports. Arguably, as the capital city, Helsinki was under greater pressure to build high-class sites for international competitions than other cities in Finland. The challenge of a fair division of public funds between the top-level sporting venues and those for wider group of practitioners was noted in the Social Democratic newspaper *Demari* in 1977: Here 'there is a stadium accommodating 50,000 people..., [and] the biggest ice-hockey hall, velodrome and indoor sports hall complex in the country... Thus, it may be surprising, that per inhabitant, Helsinki is the municipality that has the fewest ballgame areas, swimming places, and halls in Uusimaa province'.[70] Finally, although the building of sports facilities was initially also deemed to alleviate behaving problems among youth, a report by a committee investigating the youth problem in 1970 stated that the organised sports opportunities and built facilities did not seem to reach to those with unsocial behaviour.[71]

Gender and sport

Public support for competitive top-level sports as well as for large indoor ballgame halls and ice stadiums was also criticised for neglecting women and their needs. Men dominated participation in competitive sports, as well as in the administration and planning for sports.[72] According to the 1976 City Statistics of Helsinki, men were more active in competitive sports than women: participation rates were 18 per cent and 4 per cent respectively. However, of those surveyed, women (67 per cent) were more active in more

---

67 *Liikuntapolitiikan linjat 1990-luvulla*. Liikuntakomitea. Komiteanmietintö 1990:24, Liite: eriävä mielipide, Kimmo Suomi, 2.
68 *Liikuntapolitiikan linjat 1990-luvulla*, 245.
69 K. Ilmanen, *Kunnat liikkeellä: Kunnallinen liikuntahallinto suomalaisen yhteiskunnan muutoksessa 1919–1994*. Jyväskylä: Jyväskylän yliopisto 1996, 167.
70 'Helsingissä pahin puute palloilu- ja uintitiloista', *Demari* 28.3.1977.
71 Helsingin kaupunginarkisto (Helsinki City Archives), Helsingin kaupunginvaltuuston asiakirjat, Helsingin kaupunginhallitus, mietintö nr. 14 – 1970, Liite A: Nuorisokomitean mietintö, 10.
72 J. Viitanen, 'Kehitysalueet liikuntasuunnittelussa heikoilla', *Stadion 1/1972*; K. Heinilä, *Nainen suomalaisessa liikuntakulttuurissa*. Jyväskylä: Jyväskylän yliopisto 1977, 7–9, 95–115.

informal, leisure-time sports than men (59 per cent).[73] Male dominance in sports administration went back a long way. In the 1960s, it was very rare that municipal councils in Finland would elect women to their sports boards. During the 1970s it became fairly normal to have at least one woman on the board. In Helsinki, there were three women out of nine members on the Sports Board in 1954 and during the subsequent decades two women were the norm.[74] Kirsi Saarikangas, who has widely written on women's experiences in the forest suburbs of Helsinki, has argued that built outdoor sports spaces, such as ice-hockey rinks, were 'men's space' in suburban life. Sports fields represented one of the few masculine places where men could be 'true actors'. This was especially the case in the 1950s and 1960s when suburban living was dominated by mothers and children with their everyday pursuits focussed around the home, the shopping centre and the children's playground.[75]

In the early 1980s the uneven provision of facilities led sports scientists in Finland as elsewhere to question whether the ideal of 'sport for all' was actually being realised for women, elderly and other marginalised groups in the sports sector.[76] The result was a call for a more creative approach in sports site planning, emphasising enjoyment and relaxation more than rationality and competitiveness.[77]

In terms of gender equality in sports, the 1990s was an important decade. Nationally, a balance was achieved between men and women in sports board membership in the late 1990s. This followed from the wider discussion of the equality between sexes in politics and work life, as well as in the administration of sports.[78] In Helsinki, the share of female members in the Sports Board reached nearly a half in 1997, and in 2005, five out of the nine Board members were women. However, most of the top officials in the municipal sports sector continued to be men.[79] The real significance of the better gender balance in municipal sports boards may also be questioned. It has been argued that women gain more access to an organisation or institution

---

73 *Tilastollisia kuukausitietoja Helsingissä 1976*. Helsingin kaupungin tilastotoimisto 1976, 217. According to the report, 29 per cent of women and 23 per cent of men were not active in any type of sports.
74 Ilmanen, *Ensimmäisenä liikkeellä*, 250–255.
75 K. Saarikangas, *Asunnon muodonmuutoksia: Puhtauden estetiikka ja sukupuoli modernissa arkkitehtuurissa*. Helsinki: Suomalaisen Kirjallisuuden Seura 2002, 464; Saarikangas, 'Nuoruus lähiössä', 53–55.
76 L. Heinilä, 'Nainen urheilussa', *Liikunta ja Tiede* 1/1982; P. Oja, 'Kuntoliikunta ja erityis- ryhmien liikunta: Liikunta- ja kansainvälinen yhteisymmärrys kongressi 7.–10.7. Finlandia talolla', *Liikunta ja Tiede* 5/1982.
77 T. Pyykkönen, 'Oikea aivopuolisko ja liikuntasuunnittelu: Valtakunnalliset III sporttipäivät Jyväskylässä 18.–19.4.1986', *Liikunta ja Tiede* 3/1983.
78 P. Aalto, 'Naiset, liikunta ja tasa-arvo', *Liikunta ja Tiede* 3–4/2000. The official launch to equalitarian work in sports was launched, when the Ministry of Culture in Finland, Tytti Isohookana-Asunmaa, set up a committee to examine women's role in Finnish sports culture in 1994.
79 Helsingin Liikuntaviraston arkisto (The City of Helsinki Sports Department Archives), Helsingin kaupungin liikuntalautakunnan esityslistat 1997–2005; *Miehet ja naiset liikun- nassa ja urheilussa 2004: Selvitys miesten ja naisten asemasta suomalaisessa liikuntakult- tuurissa*, SLU-julkaisusarja 2/05, yhteenveto.

when its relative power has diminished.[80] Significantly, due to the economic recession of the early 1990s, the role of the municipal sports boards was curtailed. The gender question is still a controversial topic in the sports sector in Helsinki. In 1999 *Helsingin Sanomat* noted: 'Women use the gym halls of the suburb schools, or pay high prices for private aerobics classes. Girls go to private stables for riding classes'.[81] This would imply that women are still too much on the margins of public provision and forced to pay to fulfil their sporting ambition.

Sport for all and the environment

The Sports Act was amended in 1999, and compared to that in 1980, it emphasised more clearly the role of sports in supporting 'societal development'. Values such as tolerance and equality were stressed.[82] Despite the diversified sports culture at the start of the twenty-first century and increasingly contested municipal funds, the city of Helsinki still promotes the idea of offering sport for all. In the Policy Guidelines for Sports in Helsinki 2001–2010, approved by the City Council in 2001, it was acknowledged that 'accommodating exercising and sports is part of the municipality's basic level services'. The document however recognised that 'the public resources for sports have not been equally targeted at different sports and sports practitioners. Women use private sport services twice as much as men.; ... men also take part more in sports club activities than women and thus receive more financial support from the city for their organised sporting practises.'[83] Nevertheless, it is worth mentioning that organised sports play an important role for children and youngsters in Helsinki. In 2001, 47 per cent of children of ages between 3 and 19 years participated in sports club activities in Helsinki. Sports traditionally regarded as 'boys games' such as floorball and football are also increasingly popular among younger girls in Helsinki and in Finland in general.

Although the idea of providing neighbourhood-level playing spaces is not anything new, it seems, there has been a recent wave of interest in the issue from various public authorities, for example the Ministry of Education and the city of Helsinki.[84] There is a consensus that people are more active, when their immediate living surroundings offer possibilities for exercising, such as safe bicycle paths, lit recreation routes, easily accessible sports

---

80 L. Kolbe, 'Helsinki kasvaa suurkaupungiksi – julkisuus, politiikka, hallinto ja kansalaiset 1945–2000', in L. Kolbe and H. Helin, *Helsingin historia vuodesta 1945:3*. Helsinki: Helsingin kaupunki 2002, 478–479.
81 A.L. Pyykkönen, 'Liikuntapaikat rytinällä kuntoon', *Helsingin Sanomat* 26.10.1999.
82 S. Paavola, 'Liikunnan päätöksenteon eettisiä lähtökohtia', in K. Ilmanen (ed.), *Pelit ja kentät: Kirjoituksia liikunnasta ja urheilusta*. Liikunnan sosiaalitieteiden laitos, tutkimuksia 3/2004, 92–93; 'Liikuntalain uudistus korostaa tasa-arvoa', *Helsingin Sanomat* 12.3.1997.
83 *Helsingin liikuntapoliittinen ohjelma vuosiksi 2001–2010*, Helsingin kaupungin liikuntaviraston julkaisuja A4, 2001, 19.
84 *Liikuntapaikkarakentamisen suunta 2004*. Opetusministeriö: Valtion liikuntaneuvosto 2001, 6; Helsingin liikuntapoliittinen ohjelma vuosiksi 2001–2010, 19.

grounds. Children, the elderly and all those who cannot access centralised sports complexes by private cars benefit from nearby exercise possibilities. Promoting neighbourhood facilities or nearby natural spaces is also seen from the environmental protection point of view. Large box-like sports halls have rarely succeeded in blending into their surrounding environment, and often consume a lot of energy, and create pollution from traffic, if they cannot be accessed by public transport. In the 1990s, the Finnish subsidy system for sports sites was criticised for supporting large indoor sports facilities rather than small-scale developments and exercise opportunities close to nature. As in other sectors, the notion of sustainable development was connected to sports in the late 1990s; and the Sports Act of 1999 required that the sports sector should promote sustainable development, and noted the importance of close to nature sports, as well as small-scale sports site developments close to residential areas.[85]

However in spite of efforts by public policy makers to emphasise sustainable activities in sports, there is also a counter-trend in the development of sports spaces. Various sports, traditionally practised outdoors, have moved indoors.[86] Since 1998 five indoor skiing tunnels have been built in Finland, and in 2009 the first facility of this kind will be opened in Helsinki. Furthermore, another skiing tunnel as well as an indoor down-hill skiing centre is projected for Espoo, but the economic downturn starting in 2008 has delayed these projects. Football is another sport, in which practitioners increasingly prefer to play in indoor halls or on heated grass lawns. The development of these artificial sports spaces has sprung from an ideology to provide competitive sports, as well as individual sports enthusiasts, with the optimal practising environment through all the seasons, that is, a place where lighting and temperature are predictable.[87] In addition, commercialised sports providers have sought to create spaces with maximum comfort to attract customers. Critics have been worried that this is leading to people's alienation from nature and its seasonal changes. From the environmental standpoint, people should adapt to seasonal changes instead of attempting to mitigate them. In 2007, the head of the Finnish Environment Institute, Lea Kauppi, called for new innovative ways to arrange leisure and sports services in an era of climate warming. For instance, she suggested that instead of using energy to cool down artificial skating rinks in Helsinki, they should be transformed into roller skating rinks.[88]

---

85 Paavola, 'Liikunnan päätöksenteon eettisiä lähtökohtia', 92–93; K. Wuolio, 'Urheilurakentamisen tuki puntarissa: Liikuntapolitiikan ja elinkeinopolitiikan rajanveto vaikeaa', *Liikunta ja Tiede* 1/1999.
86 H. Herva, 'Liikuntaympäristön kehittäminen', in T. Haukilahti and M. Ilmarinen (eds.), *Liikunnan yhteiskunnallinen perustelut: Tieteellinen katsaus*. Jyväskylä: LIKES 1994, 307.
87 U. Hämäläinen, 'Juhannushiihtoa Helsingissä', *Helsingin Sanomat* 19.6.2009; S. Korkman, 'Hiihtoputki saattaa saada lähtölaukauksen jo marraskuussa', *Länsiväylä* 27.8.2009; H. Eriksson, 'Pääjohtaja Kauppi: Sopeutumiskeinot vain kiihdyttävät ilmastonmuutosta?', *Helsingin Sanomat* 3.9.2007; Bale, *Landscapes of Modern Sport*, 109–110; E. Vasara, 'Toiminnan ja ohjauksen kilpajuoksu', in Pyykkönen (ed.), *Suomi uskoi urheiluun*, 390.
88 H. Eriksson, 'Pääjohtaja Kauppi: Sopeutumiskeinot vain kiihdyttävät ilmastonmuutosta'; H. Hautajärvi, 'Hiihtämään helteellä!', *Arkkitehti* 5/2006, 18–19.

Conclusion

This chapter has explored the development of sports sites and the concept of 'sport for all' in late twentieth-century Helsinki. We have seen how there has been an increasing trend towards indoor sports, but at the same time outdoor sports have remained important or have developed in new directions. The most popular sports sites for adult residents in Helsinki remain pedestrian ways and recreation routes.[89] The popularity of these sports underlines the fact that in providing meaningful sites for sports, local authorities need to consider the entire urban structure, not only separate sports facilities.[90] Maintaining the idea of green corridors and connecting recreation routes in Helsinki supports the idea of exercising close by the nature, which can also be regarded as environmentally friendly and offering possibilities for different people with different sporting needs. As this chapter has indicated, despite the adoption of a 'sport for all' ideal in the 1960s, questions about the right balance between different types and levels of sport and between 'masculine' and 'feminine' sports continue to be negotiated. Also, the commercialisation of sports and the increasing demands on resources have brought new challenges to offering sporting possibilities for all.

---

89 *Helsinki liikkuu 2005–2006 tutkimus*. Helsinki: Helsingin kaupungin liikuntavirasto 2006, 35. Survey consisted of Helsinki dwellers aged 19–65 (N=549). The survey also included a survey on children aged 3–18 (N=440), whose most used sports sites were built outdoor facilities and ballgame halls.

90 For a recent discussion on combining better land use planning and sports sites see V. Rajaniemi, *Liikuntapaikkarakentaminen ja maankäytön suunnittelu: Tutkimus eri väestöryhmät tasapuolisesti huomioon ottavasta liikuntapaikkasuunnittelusta ja sen kytkemisestä maankäyttö- ja rakennuslain mukaiseen kaavoitukseen*. Studies in sport, physical education and health 109, Jyväskylä: Jyväskylän yliopisto 2005, 171.

PIM KOOIJ

4 Urban Green Space and Sport: The Case of the Netherlands, 1800–2000

The relationship between green space and sport was established in the last quarter of the nineteenth century. Industrialisation and urbanisation were the main instigators. Thanks to industrialisation, which generated higher incomes for some groups but also more time due to the eight hours movement of industrial unions, a boom in sports associations became possible. In addition, industrialisation stimulated the growth of Dutch cities. Defortification, which was a consequence of this urban growth, enabled the redesign of Dutch cities and was characterised, among others things, by increased attention to urban green spaces. Parks and sports accommodation were combined at the outset, but in the late twentieth century segregation became more common.

Urban green areas before 1800

Since the middle ages almost all Dutch cities had been walled. Within these walls there was little room for green spaces. Nevertheless, they were not completely absent. Farms and extended gardens could be found in every medieval city, which proved important when the cities were under siege. However, public green space was largely absent. Only in a few cities like Nijmegen (Valkhof) and Maastricht (Vrijthof) could some public green space be found. In addition, the Hague had no walls and was therefore able to contain some green avenues. Convent gardens and graveyards were not public places but became so after the Reformation in the sixteenth century. This led to an important extension of urban green areas.

In the early seventeenth century, particularly during the twelve-year truce with Spain, many cities were enlarged. In the new parts of the cities green nurseries anticipated the building of houses, but in some cities, like Groningen, this never took place, or at least not for centuries. These extensions therefore retained their green character.

Nonetheless, even in medieval times there were some more or less public green areas, but they were located outside the walls. Examples are the urban forests like the Haarlemmerhout near the city of Haarlem and the Mastbos near Breda. They originated as lords' hunting grounds, but over the centuries they gradually came under municipal jurisdiction. In early modern times

Figure 4.1 Maliebaan Utrecht 1713. *(De Utrechtse Archieven).*

they were used as sources of wood and for recreation.[1] There were many inns and taverns situated there, some of ill repute.

Furthermore, the walls of every city served as promenade space. In a number of cities avenues of trees were situated there to make the Sunday stroll more pleasant. This was, for instance, the case in the city of Leeuwarden after 1639.[2] In Utrecht the Maliebaan, constructed in 1637 in a new part of the city, served the same purpose. Maliebaan means a place to play pall mall (maille).[3] This was the first time that public green places for sports activities were created. They were then constructed in other cities, such as the Hague. In Amsterdam in the seventeenth century, an area left-over after the building of the canals was given over to gardens.[4] It could not be built upon because of the water level. The gardens in this 'Plantage' were rented to individual inhabitants, but together they formed a pleasant green area. A botanical garden belonging to the university was also located there. These gardens could also be found in the other university cities: Leiden, Groningen and Utrecht. In Utrecht in the eighteenth century a small urban wood was planted in a sparsely populated part of the city, in the form of a star (Sterrebos), while in other cities, Groningen for instance, similar woods were located outside the walls.

1 Forests were also cut down by enemies to make wooden siege engines.
2 M. van Rooijen, *De wortels van het stedelijk groen.* S.l. 1990, 37.
3 The Utrecht Maliebaan was bordered by eight parallel rows of trees. T. Wilmer, *Historisch groen: Tuinen en parken in de stad Utrecht.* Utrecht: Matrijs 1999, 17. In 1768 the Maliebaan was extended with a park and a pond. The Maliebaan was also suitable for horse racing.
4 E.R. Taverne, *In 't land van belofte in de nieuwe stad: Ideaal en werkelijkheid van de stadsuitleg in de Republiek 1580–1680.* Maarssen: Schwartz 1978.

One form of green space missing in the Netherlands was a royal park in the capital. The Netherlands had no royal family before 1806. The family of Orange-Nassau, who were stadtholders in most provinces, kept their modest palaces outside the cities (Honselaersdijk for Frederik Hendrik, 't Loo and Soestdijk for William III). Only the Valkenberg park in Breda, the cradle of the Orange family, became a public green space.[5]

Urban green areas after 1800

In the nineteenth century the shape of Dutch cities changed radically. This started in the west of the country. Thanks to the construction of the Nieuwe Hollandsche Waterlinie from 1813, the ramparts of the cities enjoying its protection could be removed. The Waterlinie was a defence system of canals and locks by which lower-lying land could be inundated. A set of fortifications was built on strategic higher ground, particularly around the cities of Amsterdam and Utrecht.

The extension of the cities in the west was controlled by members of the elite, who dominated the municipal councils. They wanted wide boulevards, surrounded by their houses, public buildings and parks, to be able to continue their promenades. Frequently, the Vienna Ringstrasse was their ideal, but most cities were unable to generate enough capital to realise these aspirations. At Amsterdam in 1864, for instance, plans were laid for a very expensive belt in the south of the city including the Amstel hotel and the Amsterdam 'Crystal Palace' – the Paleis voor Volksvlijt. Only a small part was ever completed, while in other areas housing was built for labourers and factories. Nevertheless, the part realised included two parks, the Vondelpark and the Sarphatipark.

Parks at that time were judged important. In the eyes of the elite and middle classes industrialisation and the related growth of cities had generated unhealthy stone deserts. Rapid urbanisation also threatened the countryside, which was vanishing fast. As a result, a new vision of nature arose, put forward by nature propagandists like the teachers Jac. P. Thijsse and E. Heimans, who edited the journal *De Levende Natuur* (the living nature). They developed a new environmental vision by trying to increase the public understanding of nature. They directed expeditions to natural, unspoiled parts of the country and initiated the establishment of a National Trust, *Natuurmonumenten*. One of the first acquisitions was a lake near Amsterdam, where a refuse dump had been planned. Some country sites followed, which at that time were seen as having natural features by many nature conservationists. However, they were, for the greater part, man-made in accordance with the Arcadian-paradisiacal vision of nature, which had emerged in Britain in the sixteenth century and on the continent somewhat later.[6] This vision of nature preferred a civilised, clean nature, which was not too wild. The English Landscape

---

5 van Rooijen, *Wortels*, 62–55.
6 The third vision on nature I distinguish is the utilitarian one, which had already taken shape in 10,000 BC and which implies that nature is shaped to meet humanity's needs. P. Kooij, *Mythen van de groene ruimte*. Wageningen: Wageningen University 1999.

style fits this vision. This was also the style of most nineteenth-century city parks. These parks, however, were sometimes so 'wild' that they could also be seen as natural.

Viewed this way, city parks could play a role in meeting the need for nature, greenery and health which emerged alongside industrialisation. Therefore, city parks became a necessary part of the expansion of every city. Famous families of landscape gardeners like Springer and Zocher designed these parks and so there is considerable uniformity. They even redesigned the remaining urban forests. When a law of 1874 permitted cities in other parts of the country to remove their fortifications, the example of the cities in the west and centre of the country was imitated.

In addition to these parks, villa parks for the well-to-do were created. The ones in the Hague were particularly popular. The greater part of the Rotterdam elite moved to the Hague. The social counterpart of the elite villa parks were the parks created by industrialists for their labourers.[7] One of the first was the Agneta Park in Delft, created by J. C. van Marken, a producer of alcohol and yeast, and named after his wife. Construction started in 1885. The Van Markens lived in the middle of the park in the villa Rust Roest (rest rusts). Not every labourer appreciated this social control. There were no sports facilities in this village except a building with a billiards table and a skittle-alley.

In fact, the Agneta Park was the only nineteenth-century park for labourers. There were some other industrialists who built for their workers, but only streets or individual houses. The boom in these industrial villages was in the early twentieth century, when dozens were built, many of them according to the garden city concept. There were concentrations near Utrecht (founded by metal factories), Rotterdam (shipbuilders), the province of Limburg (mines), and Amsterdam (a cooperative village named Betondorp 'concrete village'). Famous examples are the Philips village in Eindhoven, which was constructed from 1909 on for employees of the Philips plant and which grew to 1,000 houses, and 't Lansink in Hengelo, built by the Stork company. Both parks had sports facilities. It is appropriate at this point to consider where the need for such facilities originated.

The start of organised sports in the Netherlands

Modern sports culture in the Netherlands is rooted in the second half of the nineteenth century. This does not mean that before that time sports were completely absent. In the middle ages and early modern times there were many sporting activities linked with hunting and warfare like archery, horse riding, and fencing. The citizens militia had special drilling grounds in the towns, called Doelen.[8] In addition to these militarily inspired sports, there were also games of fives, skittles and pall mall, which had more of an entertainment value. The first fives court is mentioned in 1371. It was in a

---

7 H.J. Korthals Altes, *Tuinsteden tussen utopie en realiteit.* Bussum: Thoth, 2004.
8 C.L. Verkerk, 'Sport en spel in de Middeleeuwen', *Groniek* 144 (1999), 265–277.

convent garden in Utrecht.[9] The nobility and the middle classes played indoors. The lower social classes played in the open air, for instance near inns in the countryside. It is notable that fives became one of the first organised sports in the Netherlands. It was introduced in the northern province of Friesland by dyke builders from the province of Holland and became very popular in that province. In 1853 shopkeepers from Franeker formed a permanent commission, the PC, to organise an annual fives competition at a special field near an inn. This competition continues to this day at the same place.

Another sport dating from the middle ages is skating. Skating was initially a means of travel when the canals were frozen and the barges – a sixteenth-century innovation – could not move. However, skating also implied a form of social mingling in which the lower classes mixed fairly easily with the upper ones, as did men with women.[10] The Hofvijver, a pond in the centre of the Hague, near the parliamentary buildings, was such a meeting place, as were the frozen rivers in Amsterdam and Rotterdam, where so-called ice fairs were held. At that time skating was not a competitive sport and speed skating only started around the middle of the nineteenth century. The sport became organised quite early on. The first skating club was founded in 1840 in the Frisian town of Dokkum, and in 1882 a national skating association was founded.

This national association for skating was paralleled by a number of associations for other sports, which were founded more or less at the same time.[11]

Table 4.1 Dutch sports organisations (dates of establishment)

Year	Sport	Year	Sport
1847	sailing and rowing	1901	water polo
1868	gymnastics	1903	korfball
1874	hunting	1903	physical strength competition
1879	horse running	1904	motor racing
1882	skating	1908	fencing
1883	cricket	1909	skittles
1883	cycling	1909	walking
1883	rowing	1911	boxing
1885	pall mall	1911	billiards
1888	swimming	1912	baseball
1889	football + athletics	1914	golf
1890	shooting	1925	badminton
1897	fives	1927	handball
1898	racing	1927	skiing
1899	tennis	1928	bobsleigh

9 P. Breuker, 'Kaatsen in international verband: beperkt grensverkeer mogelijk', in P. Breuker and W. Joustra (eds.), *Sporthistorie tussen feit en mythe*. S.l. 2004, 101–133, 103.
10 J.H. Furnée, 'The thrill of frozen water: class, gender and ice-skating in the Netherlands, 1600–1900', in S. C. Anderson and B. H. Tabb (eds.), *Water, Leisure and Culture*. Oxford, New York: Berg 2002, 53–69.
11 H. Dona, *Sport en socialisme: De geschiedenis van de Nederlandse arbeiderssportbond 1926–1941*. Amsterdam: Van Gennep 1981, 122.

A leading promoter of organised sport was Pim Mulier from Haarlem. He attended school at Ramsgate in England and learned there how to play soccer. In 1879 he returned and set up the first football club of the Netherlands – *Haarlemsche Football Club* (H.F.C.). Their field was a meadow north of the Haarlemmerhout. It was called Koekamp (cow meadow). Other sports were also practised there. Another pioneer was Jan Bernard van Heek from Enschede, son of a cotton mill owner, who was an apprentice in England, from where he brought football to his home town.

In 1889 Mulier was one of the founding fathers of the *Dutch Football Association*. He also introduced handball, hockey and rugby, organised the first athletics competition in the Netherlands and propagated tennis. In 1890 he skated the first Elfstedentocht along eleven Frisian cities, which is today the most famous long-distance skating competition in the Netherlands.[12] In 1898 he ensured that football, athletics and fives became part of military training.[13]

In the nineteenth century sport was primarily practised by members of the elite and middle classes. In some sports, such as tennis and hockey, there was an effective black-balling of other classes, and in football certain clubs likewise kept their exclusivity. An interesting case is the social segregation which eventually occurred in skating. In the Hague, the elite, detesting the new form of high speed skating with women participating in their underwear, founded an exclusive skating club. For this they obtained permission to have exclusive use of the ponds of the estate Zorgvliet, owned by the king's sister.[14]

Around 1900 the first clubs for the lower classes were founded. In 1897 students of the Haarlem technical school founded *Haarlem*. Clubs especially for labourers were also established in other cities, for instance *Blauw Wit* in Amsterdam. Korfball was developed particularly for people from the lower classes, but in most sports, associations remained in the hands of the elite. In gymnastics and swimming special divisions for the lower classes were created, and in cycling the organisation was split into one part for labourers and lower-middle classes, and one for the well-to-do.[15]

In the meantime the number of organised sportsmen almost doubled every decade, from 30,000 in 1900 to 65,000 in 1910, 150,000 in 1920, 278,000 in 1930 and 490,000 in 1940. The fastest growing sport was football, the number of players rising between 1910 and 1920 from 7,500 to 48,000.[16] This was primarily the result of the mobilisation of the Dutch army during the First World War. The Netherlands remained neutral so there was plenty of free time for the troops, whilst the English soldiers detained by the Dutch made popular playing partners. In this way members of the lower classes became acquainted with football. This led in 1926 to the foundation of the

---

12 R. Stokvis, 'Pim Mulier en de sport in Nederland', in H. Beliën, M. Bossenbroek and G.J. van Setten (eds.), *In de vaart der volkeren, Nederlanders rond 1900*. Amsterdam: Bert Bakker 1998, 289–298.
13 Breuker, 'Kaatsen', 108.
14 Furnée, 'The thrill of frozen water'.
15 Dona, *Sport en socialisme*, 124–125.
16 Dona, *Sport en socialisme*, 125–126.

Nederlandsche Arbeiders Sportbond (Dutch Labourers' Sport Association). According to De Rooy, this was a very interesting development because whilst members of the association reacted against the bourgeois sporting culture, they took over notions of fair play and the Olympian ideal that participation is more important than winning.[17] At the same time, Protestant and Roman Catholic clubs were also founded as an expression of the cleavage in Dutch society at that time. Each church had its own association, the Roman Catholic one dating from 1915 and the Protestant association founded in 1929.[18]

The Roman Catholic and Protestant clubs consisted for the greater part of people from the middle classes. Just like the labourers, and unlike the elite who dominated sports previously, they needed help finding accommodation. Therefore, local politics became involved. The introduction of general suffrage for men in 1917 and women in 1919 fundamentally changed the composition of the municipal councils. The liberal elite was now challenged by elected representatives from Protestant, Roman Catholic and socialist parties. The socialist members of the council were particularly vocal in their efforts to promote sports activities.

Sport, parks and sports parks

As long as sport was an activity of a select group of urban citizens, finding a place to practice was not a problem. They hired a field from the mayor of Haarlem, as we noted above, or rented some land from a farmer just outside the city. But as the number of organised sportsmen grew, local government and some industrialists started to organise opportunities for sport more systematically. This happened in four ways: first, the creation of sports grounds in existing parks; second, the construction of new parks including sports grounds; third, the incorporation of sports accommodation in garden villages linked to industry; fourth, the foundation of areas dedicated exclusively to sports activities.

The creation of sports grounds in existing parks took place step by step. The older landscaped parks generally had some large lawns. Initially, visitors to these parks were not allowed to walk on the grass, but over the course of time this was tolerated and subsequently some lawns were transformed into football fields or tennis courts. New parks were constructed as a more effective way to offer sports accommodation. In the early twentieth century the phenomenon of the people's park (Volkspark) arose. This concept was inspired by developments in Great Britain and Germany, both countries with much longer industrial traditions than the Netherlands. Rapid urbanisation needed to be counterbalanced by the creation of healthy green areas where fresh air could be found and where the walking middle classes could set a civilised example to the lower classes. From about 1840, British parks were

---

17 P. de Rooy, 'De wijsheid van de jongensclub', *Groniek* 144 (1999), 283–295.
18 C. Miermans, *Voetbal in Nederland: Maatschappelijke en sportieve aspecten.* Assen: Van Gorcum 1955. In 1940 they all merged with the national organization.

often created by municipal authorities using grants of money and land from the elite. In Germany, from about 1860, the local administration had a more decisive role.

The first park in this tradition appeared in 1872 in the cotton city of Enschede where the heirs of the industrialist H. J. van Heek presented a park to the community to be used by labourers. This was very early: in other cities in the non-western regions of the country parks were being constructed in spaces left after the clearance of fortifications. As noted earlier, these parks were initially designed for the upper and middle classes. At first there was little difference in shape between the upper and lower class parks. But in 1918 a new donation from the van Heek family followed. This time a playground for children and a sports field were incorporated. In Rotterdam the Kralingse Bos was constructed in the annexed municipality of Kralingen between 1920 and 1927 in the style of a Volkspark, with accommodation for sports, a big lake for sailing and canoeing, and tracks for cycling and horse riding.

The Housing Act (1901), which obliged every larger municipality to design an expansion plan, was also a stimulus for the construction of parks. This inspired urban civil society to define a complete set of urban functions, in which sport, leisure and recreation were incorporated. A number of these expansion plans were designed by the famous architect H. P. Berlage, who always included parks and sports accommodation in his plans.

Playgrounds and sports fields could also be found in almost every garden village created after 1900. A football pitch, for example, was incorporated in the Philips park in Eindhoven. This is where the *Philips Sport Vereeniging* (PSV) started to play, and still plays, though it is now in an enormous stadium which totally dominates the Philips village. More football clubs started life in garden communities, for instance Elinkwijk in Utrecht. 't Lansink, built by Stork in Hengelo, is one of several with an open air swimming pool. Separate sports complexes were created from the 1930s. A major source of inspiration was the complex which was constructed for the Olympic Games in Amsterdam in 1928. As will be seen below, socialist aldermen played an important role in providing these municipal services. The inner courts of houses built by housing associations were also sometimes used for sports, especially korfball clubs.

The four ways in which sports accommodation could be created in all Dutch cities worked out at different intensities and mixes, depending for instance on the size of the city, the economic structure, the spatial structure, and the political constellation. I will therefore now turn to four individual cities which arguably are representative of many others. Specifically, these are the two metropolitan cities – Amsterdam and the Hague, of which Amsterdam had been walled while the Hague had not – and two second rank cities – Groningen, an industrial and service city in the north of the Netherlands, and Zwolle, situated in the centre, a city which never industrialised but became a service centre. After having explored the different local origins of sport and the green areas, the general theme will be continued by indicating how subsidies from central government, the emergence of a national planning tradition, and the demonstration effect of some cities resulted in a growing uniformity in the relationship between sport and green areas after 1945.

Amsterdam

As already mentioned, in nineteenth century Amsterdam, instead of a green belt, only some rather isolated green spaces were available. The Plantage, thanks to new draining techniques, was built upon after 1857. As a result, the Amsterdam elite lost this area for recreation, but thanks to the presence of the Hortus Botanicus, a small park dating from 1812, and the zoological garden Artis, some green places remained.[19] The Vondelpark was an initiative in 1864 of Amsterdam's top elite. It was financed by donations and the selling of adjacent lots for the building of villas. The size of the park was 10 hectares and it was designed by J. D. and L. P. Zocher.[20] The Sarphati Park was named after the doctor, industrialist and city patron Samuel Sarphati and was much smaller than the Vondelpark, 4.21 hectares. The land was gifted by Sarphati to the municipality and was opened in 1885.

At that time the plan for a major extension of Amsterdam, primarily meant for the elite, had given way to a more moderate plan, drawn up by J. Kalff, who also made room for houses for the middle classes and even labourers. At the fringes of the belt, on space not suitable for the construction of houses, he designed two parks – the Oosterpark (East park) and the Westerpark (West park). The Westerpark plan was initially for 10 hectares but a gas factory was also built on the same spot, which left only 5.63 hectares. The park was constructed in the 1890s according to plans by the municipal engineer J. G. van Niftrik. He also drew up the first plans for the Oosterpark, but here the final construction was by L. Springer.

The Sarphati Park was too small to have special facilities but the Vondelpark and the Oosterpark were described as walking and driving parks. They had different networks of lanes for pedestrians and coaches. The latter were soon taken over by cyclists. The Oosterpark and Westerpark were the first nineteeth-century parks with planned places for sport, including a lawn specially intended for cricket. In fact, this did not require much adaptation. The same was the case for a lane in the Oosterpark, 500 meters long, for harness races. The Vondelpark became the home of tennis clubs, like *Festina*.[21]

Around 1900 there was a lack of accommodation for organised sport in Amsterdam. As a result, many clubs led a wandering existence. *Ajax*, for instance, played on some meadows in Amsterdam north but moved in 1907 to Watergraafsmeer in the south and built its first stadium there in 1911. The football clubs related to the big shipyards in Amsterdam, like *De Volewijckers*, usually had their fields near the enterprises.

The Olympic Games in 1928 were a major stimulus for sport in Amsterdam. A complex was built for this occasion which replaced an older

---

19 M. Wagenaar, *Amsterdam 1876–1914: Economisch herstel, ruimtelijke expansie en de veranderende ordening van het stedelijk grondgebruik*. Amsterdam: University of Amsterdam 1990.
20 In 1954 it was donated to Amsterdam.
21 R. Stokvis, 'De opkomst van de moderne sport', in P. de Rooy (ed.), *Geschiedenis van Amsterdam: Tweestrijd om de hoofdstad 1900–2000*. Amsterdam: SUN 2007, 138–149.

one from 1914. It consisted of a swimming pool, a fencing stadium, another for strength sports, and the Olympic stadium with 40,000 seats for football, athletics and cycling. After the games the stadium was used by football clubs like *Blauw Wit* and *F.C. Amsterdam*, while *Ajax* (before the opening of the Arena in 1996) played its important matches there. In the Olympic year of 1928 it was decided to create the Amsterdam forest (Amsterdamse bos). The main inspiration was not the euphoria over the six Dutch gold medals at the games, rather it was due to measures to reduce unemployment combined with a growing interest in nature. Jac. P. Thijsse, a leading environmentalist in the Netherlands, was one of the initiators of the plan.

The plan was made public in 1928 and implied the transformation of a polder landscape into a 900 hectare forest park. The main architect was the young Cornelis van Eesteren, who would later become a famous modernist. However, the park was designed in a robust English landscape style, perhaps thanks to the involvement of specialists like E. D. van Dissel, the director of the Dutch Forestry Commission (Staatsbosbeheer), the soil specialist D. J. Hissink, and of course Jac. P. Thijsse. Inspiration came from the Bois de Boulogne in Paris, Hampton Court near London and the Treptowpark in Berlin. An important innovation was that visitors were encouraged to walk and play on the lawns. There were even movable goalposts available for soccer, hockey, korfball and handball.

Top of the sports facilities was a rowing course, the Bosbaan, 2200 meters long and 65 meters wide. Construction started in 1934 and it opened in 1937.[22] It was also used for fishing and skating. For swimming there were many ponds, shallow and suitable for small children, while other ponds were used for canoeing. There was no real swimming pool. A sports complex was constructed for organised sports, especially for hockey, while the hockey fields in summer could be used for cricket. In 1947 there were also eleven tennis courts, four football fields and an equestrian centre. Football facilities were insufficient and there were no spaces for organised athletics. Many people nonetheless played football in informal teams or went running on their own. Therefore, it can be maintained that the Amsterdamse bos was the main place in the city for informal, unorganised sports, not least because of the extensive network of tracks for cycling and horse riding.

Organised sport was for the greater part not located in parks. There were some separate complexes, the most important being around the Olympic stadium. The creation of separate green places for sports was made an official municipal policy in the General Expansion Plan of 1934, also designed by Cornelis van Eesteren. In this plan three functions were defined – living, working and recreation. A fourth function, traffic, linked those functions which were located in separate places at short distances from each other. In fact, living quarters were separated from working quarters by recreational areas where allotment gardens and sports complexes were planned. The socialist aldermen agitated for a fast realisation of the plan. As a result, especially in the west, expansion took place around the newly created Sloterplas, an artificial lake.

---

22 Later it was widened twice, to 72 and 92 meters respectively

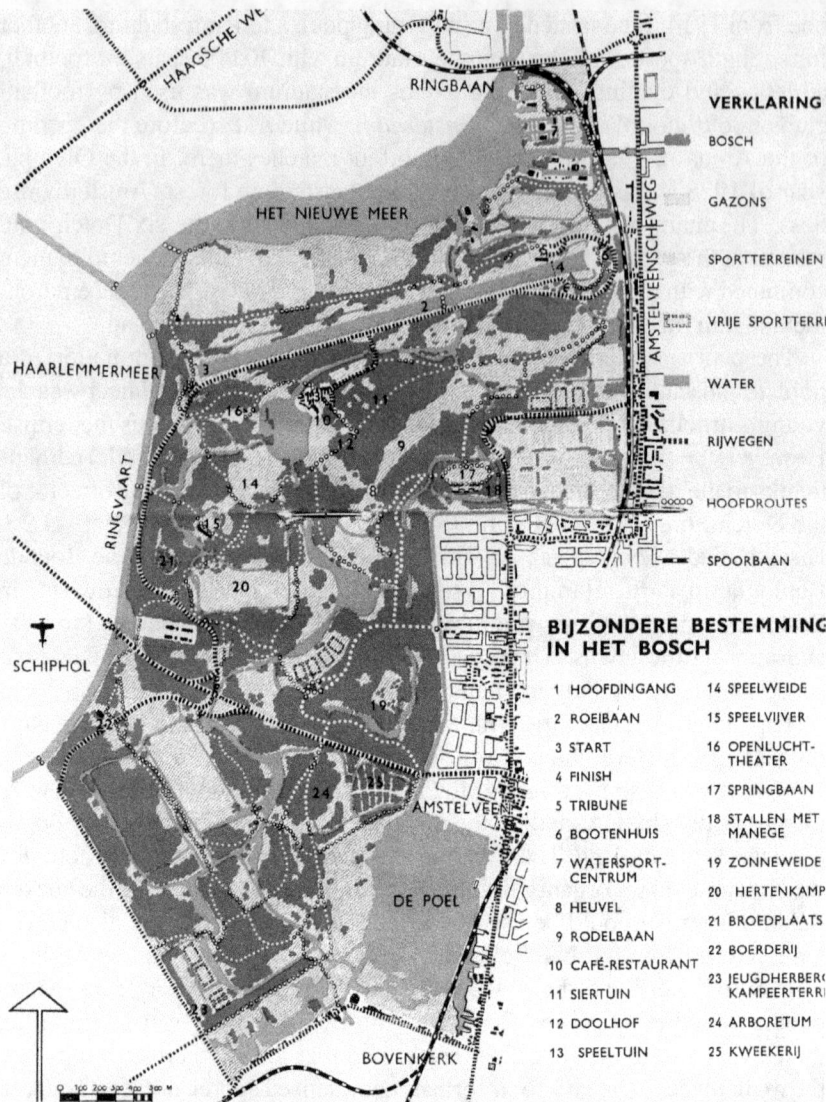

Figure 4.2 Plan for the Amsterdam Forest, 1934. (Bosplan 1934: Catalogue of the exposition).

After the Second World War the General Expansion Plan continued to guide how the expansion of Amsterdam was managed.[23] Within the recreation function there was a further differentiation, as parks and sports accommodation became completely separated. The new quarters of Amsterdam, such as Noord, Nieuw Zuid, Buitenveldert, Bijlmermeer, all got parks and sports accommodation, which for the greater part were located along the traffic axes. This pattern provided the city with a set of green corridors.

23 V.T. van Rossem, *Het algemeen uitbreidingsplan van Amsterdam: Geschiedenis en ontwerp.* Rotterdam/The Hague: NAi 1993.

Some larger parks were constructed between the individual quarters. In the Buitenveldert quarter, for instance, a central green line is formed by the Gijsbrecht van Aemstelpark, while the Amstelpark in the east marks the border with the Rivierenwijk. This park is surrounded by sports accommodation. In fact, the tradition of separate sports accommodation set by the Olympic stadium and the Ajax complex was continued, though they were now situated in a green context. The Arena displayed the opposite development. This stadium is surrounded by shops, offices and other high urban elements and has served many times as the location for pop concerts. The same is the case with the Olympic stadium, which was saved from demolition and restored between 1992 and 2000.

The Hague

In fact, the Hague could be described as two different cities – a luxurious residential city in the sandy north and north west characterised by large country estates and open green spaces, and an industrial labourer city in the south on the more unhealthy peat soil.[24] In this southern part of the city there were no green areas. These could be found in the sandy part and in the adjacent fishing village of Scheveningen. In 1613 the Haagse bos (The Hague forest), with a Maliebaan, was opened to the public, while in 1618 a network of canals, bordered with trees, was constructed. North of the Hofvijver a wide avenue, the Voorhout, was constructed. In the nineteenth and twentieth centuries most country estates were sold to real estate developers and transformed into green villa parks.[25] For most inhabitants this meant an extension of the green area, because the country estates had been strictly private. Famous architects, including members of the Zocher family, secured a mix of green [space] and red [brick].

The first villa park, the Willemspark, was created between 1855 and 1858 in a park north of the centre, which had belonged to King William II. Another property of the royal family, the Zorgvliet estate, also came into the hands of real estate developers. Parts were sold off in 1876 and during the early twentieth century. In this enormous area the Peace Palace (1913) and the Municipal Museum were built, as well as more than a hundred villas. The core of the old estate remained a park around the manor, the Catshuis, named after its creator (1643), the Dutch poet and politician Jacob Cats.[26]

In the early twentieth century the city of the Hague became an important buyer of country estates. The city even had a construction company of its own. Some of the estates were partially built over but others, like the enormous Ockenburgh estate, which was bought in 1931, were used for the

---

24 H. Schmal, *Den Haag of 's/Gravenhage De 19de eeuwse gordel, een zone gemodelleerd door zand en veen*. Utrecht: Matrijs 1995.
25 B. Koopmans, *Over bossen, parken en plantsoenen: Historisch groen in Den Haag*. The Hague: Gemeente Den Haag 2000. There were also some villa parks constructed in natural areas near Scheveningen like the Van Stolk park and the Belgische park.
26 Now it is the official residence of the Dutch prime minister.

construction of a graveyard, a campsite and a sports complex. The Marlot estate, near the Haagse bos, had already been acquired by the city in 1917. Part of it was transformed into a villa park while the rest remained a public park.

As in Amsterdam, the nineteenth-century parks had no specific sports facilities. However, the lawns in some parks were used to play cricket and later football. Around 1880 cricket was played on the Malieveld, and there were cycle routes too. Two well-known football clubs, *Haagsche Voetbal Vereeniging* (HVV) and *Houdt Braef Stant* (HBS), started here.[27] In the Scheveningse bosjes, a more or less natural area near the village of Scheveningen, a set of tennis courts, the Bataaf, was created around 1900.

In the densely populated labourers' quarters in the south of the city there were hardly any green spaces. Therefore, in 1919 the municipality started preparations for the construction of a Volkspark. The extensive Zuiderpark, designed by the municipal garden architect Westbroek and his successor Doorenbos, was opened to the public in 1931. There were an open air swimming pool, 163 metres long, and a large sports complex, especially for football, where *ADO* started to play in 1925.[28] At about the same time a counterpart to the Zuiderpark, the Westbroek park, was created in the west near the villa area. It was not a public park and had hardly any sports accommodation, but in its style it resembled the Zuider park .

Unlike Amsterdam, the Hague had no lack of green areas, and therefore no creation of green space was necessary in the greater part of the city. In fact, the opposite occurred. Green space was built upon. Even a few of the sports complexes were constructed in this way. The best examples are the Bosjes van Pex at the Sportlaan (Sport alley). Here in 1927 the meadows around a number of farms were transformed by the master-builder Willem Dudok into football pitches, tennis courts and multifunctional lawns.

The Hague did not have the chance to annex neighbouring villages, as Amsterdam did. After the Second World War, the only space left for development, apart from bombed areas, was in the south where the labourers lived. Therefore Willem Dudok, who was in charge again, had to build cheaply and at high density.[29] In the new quarters of Moerwijk, Morgenstond and Bouwlust the apartment buildings lined up in ranks, and there was no room for parks, only for some sports fields in the corner. Before building activities could reach the remaining parts of the green area, they were relocated to neighbouring municipalities, where for instance the satellite town Zoetermeer was created. To make these places attractive, large parks and sports accommodation were incorporated. As a result, the situation in the Hague remained for the greater part as it had before, with sports accommodation as enclaves in the green areas in the sandy district and in the parks in the peat area.

---

27 Koopmans, *Over bossen*, 19. In 1896 the State of the Netherlands, who owned the Malieveld, forbade this kind of sporting activity.
28 Since 1950 there has been a large stadium for *ADO*.
29 K. Stal, 'Ruimtelijke ontwikkeling', in T. de Nijs and J. Sillevis (eds), *Den Haag: Geschiedenis van de stad 3. Negentiende en twintigste eeuw.* Zwolle: Waanders 2005, 11–54.

Figure 4.3 Plan for sports facilities (tennis, football) in the Pex Forest, The Hague 1933. (Dienst Stedelijke Ontwikkeling The Hague).

In 2002 the Hague finally got expansion possibilities at the expense of its neighbours. The large quarter of Leidschenveen-Ypenburg was designed east of the city. In this area a new stadium for *ADO* was planned, surrounded by a commercial zone. It opened in 2007.

Groningen

A Dutch city of the second rank, Groningen had been a commercial centre for ages, but it began to industrialise at the end of the nineteenth century. The result was rapid growth from 39,800 inhabitants in 1875 to 74,600 in 1910. This growth was accompanied by an expansion of the city, which commenced in 1874 when the city was allowed to remove its fortifications. The Groningen elite succeeded in creating a kind of public/private Ringstrasse with boulevards in the south, university buildings in the west, a new academic hospital in the east, and a park in the north.

This park, the Noorderplantsoen, made up for the lost ability to promenade on the fortifications. The first stage was designed by the municipal architect J. G. van Beusekom, the second by the Utrecht landscape architect H. Copijn. Both preferred a romantic English landscaped style, with a bandstand and a pavilion for drinking milk.[30] At about the same time the eighteenth-century Sterrebos, which was characterised by a geometric form, was also transformed

---

30 B. Hofman et al., *Het Noorderplantsoen: Van dwingers tot park.* Groningen: Noordboek 2001.

into a landscaped park, again with a bandstand. There were no sports facilities in either park. The multifunctional lawns in the Noorderplantsoen, which in future decades would become a place for many sports and cultural activities, date from the 1920s, when the construction of the northern quarters started.

Groningen sportsmen did not have to wait long, however. Already in 1839 the teacher R. G. Rijkens organised public gymnastic exercises on a field outside the main gate.[31] In 1862 a skating association was founded, in 1878 two gymnastics clubs and in 1881 a cycle club. In 1878 the student rowing club *Aegir* was formed, followed by a general club in 1886. Organised football started in 1887, when the *Be Quick* cricket club set up a soccer team. Like many clubs, *Be Quick*, which recruited its members mainly from the elite, was an itinerant club, moving from field to field, until in 1913 they got their own ground and in 1921 even a stadium of their own.

The first official sports ground, the Noordersportterrein (Northern sports field) of 1898, was mainly used for horse racing, which was very popular in Groningen. The need for more accommodation became acute after the First World War when not only the number of sportsmen grew exponentially but also the range of sports broadened, for instance through the foundation of korfball and handball clubs. At that time, socialist members of the municipal council and socialist aldermen had secured the opportunity to influence urban politics decisively. Building for labourers had priority. The showpiece was a new quarter east of the city (Oosterparkwijk), with high quality architecture and sports facilities. A covered swimming pool was built and in the middle of the quarter a football complex was constructed with a stadium for *GVAV*, the football and athletics club for the lower classes.

In the meantime, a Volkspark was created in Groningen. It was founded by an association (1909) under the presidency of the rich industrialist Jan Evert Scholten, though the true initiator was the municipal architect J. A. Mulock Houwer. At an early stage Leonard Springer was asked to design a plan. He took the Amsterdam Vondelpark as a starting point,[32] but with much more attention paid to sport. Scholten was a keen amateur horse racer and therefore a big racetrack was constructed in the middle of the park. There was also a skating rink, while the large ponds were sometimes used for swimming demonstrations. At the corners there were football and handball fields and also a track for athletics. Tennis was played elsewhere, in a small park south of the city. Thus, Groningen could boast a very real combination of parks and organised sport.

After the war, however, this pattern was replaced by an approach which resembled that of Amsterdam. In the new quarters green belts were created with separate sports complexes were constructed at the corners. Initially, large new parks were not developed since the city did not grow that fast. However, around 2000 a big recreational park was created east of the city. Kardinge combines nature, outdoor sports facilities and indoor ones like a skating rink and a swimming pool. The new facility was accompanied by

---

31 B. Hofman, *Van Gruno tot Giro: Sportstad Groningen.* Groningen: Noordboek 2002.
32 S. Ringkvist, 'Het stadspark: van ontstaan tot uitbreiding', *Groningen toen.* Groningen: Groninger Gezinsbode 1984, 41–65.

the closing of some post-war facilities. At the same time Groningen got its own Arena, the Euroborg for *FC Groningen* incorporating a stadium, cinema, shops, fitness centres, restaurants and cafés.

Zwolle

During the twelve years truce with Spain (1609–1621), Zwolle acquired its characteristic star shape with impressive fortifications. An older part, the Koterschans, linked these fortifications with the river Ijssel. In 1828 this was transformed into a park in English landscaped style (the Engelsche Werk). This was the first green provision for Zwolle's inhabitants, which was extended between 1829 and 1848 when the fortifications were transformed into a green belt with villas and some small parks, like the Ter Pelkwijk park.[33]

These parks were too small for sports activities and the Engelsche Werk was too remote, although some football was played there. However, in the direct vicinity of the town there were sufficient meadows. Moreover, some clubs were so big and rich that they could afford to build their own accommodation. One such exclusive club was *Zwolsche Athletische Club* (ZAC), which was founded in 1893 by the local sport promoter Jasper Warner. The members practised athletics, cycling, skating, shooting, fencing and football. Warner organised and probably paid for facilities for *ZAC* in 1911. For a long time (1897–1919) he was also chairman of the *Dutch Football Association* and sometimes organised international matches in the *ZAC* stadium.[34]

Another big football club was *Prins Hendrik – Ende Despereert Niet Combinatie* (PEC) from 1910. The members of this club came from a lower social level, but could still afford their own accommodation near that of the *ZAC*. Labour clubs like the *Zwolsche Boys* or the Catholic *VIOS* and the Protestant *Be Quick '28* needed some help, however.[35] The *Zwolsche Boys* started in the Vrolijkheid complex in the neighbouring municipality of Zwollekerspel, and a municipal complex was created in 1935 for the other clubs. *PEC* also moved to that stadium but later on swapped with the *Zwolsche Boys*.

The *Zwolle tennis clubs* also had their own concrete, later on gravel, courts, while in 1934 a wooden cycledrome was constructed.[36] This shows that of the four cities described here, Zwolle was the one with exclusively separate accommodation. This pattern was continued after the Second World War. Like Amsterdam, Zwolle was expanded with green belts, bigger parks between the quarters and separate sports accommodation. The oldest made

---

33 D.J. Rouwenhorst, *Het cultureel erfgoed van Vastgoed Zwolle*. Zwolle: Vastgoed Zwolle 1999.
34 W. Coster and J. de Jong, *De Veerallee: Een rondgang door de wijk en de tijd*. Zwolle: Waanders 2004, 53.
35 J. ten Hove, *Geschiedenis van Zwolle*. Zwolle: Waanders 2005, 509.
36 A. Meijerink-Wijnbeek and F. Pfeifer, 'Honderd jaar Zwolle, de Zwollenaren en hun sport', *Als de dag van gisteren*, vol. 8. Zwolle: Waanders 1992, 179.

Figure 4.4 Interland competition (Netherlands – Belgium) in 1913 in Zwolle. (Photograph: Historisch Centrum Overijssel).

way for houses. The same happened in Zwollekerspel which was annexed by Zwolle. In 2007 the building of a new stadium for *FC Zwolle* (a continuation of *PEC*) started, modelled on the Amsterdam Arena with a lot of commercial space.

Converging traditions in the provision of urban sports facilities

The tales of the individual cities reveal different traditions in sports accommodation before 1945. While in Groningen most of the complexes were created in parks, Zwolle constructed separate accommodation. How can this difference be explained? In my opinion, this difference is primarily due to the availability or absence of general extension plans. The General Housing Act of 1901 obliged every major city (10,000 inhabitants or more) to draw up such a plan, but not every city did so. In Groningen, such a plan was almost immediately drawn up by master builder J. A. Mulock Houwer.[37] As in the older Amsterdam plans, this extension plan designed parks with sports facilities. Zwolle, however, only drew up their extension plan in 1950. The main reason for this was that the city was enclosed by neighbouring municipalities. Naturally, the socialist aldermen could not wait for parks to be planned and therefore the provision of sports facilities had a more ad hoc character. In Dordrecht, a second rank industrial city in the west, which also was enclosed and got its extension plan rather late, the same development as in Zwolle took place, also due to initiatives by socialist aldermen.

In Amsterdam, the General Extension Plan of 1934 was considered the first modern plan since the fast-growing city had experienced a series of annexations. The plan meant a shift from the ad hoc construction of sports facilities to the careful planning of sporting accommodation in recreational

---

37 It was accepted in 1906. P. Kooij, *Groningen 1870–1914*. Assen/Maastricht: Van Gorcum 1987, 214.

zones. The Hague, initially a city with a lot of green areas and ample room for extension, more or less followed the same extension pattern as other cities without fortifications, for example Helsinki, by building and creating sports facilities in forests and other green areas, but eventually this process came up against the boundaries of adjacent municipalities.

The diversity of the pre-war period was replaced by uniformity after 1945. The Amsterdam General Extension Plan became the norm for urban sports accommodation – they were constructed in green belts in extension quarters, near parks but visually separated from them. A second element, the 'petrification' of sports facilities, which started with indoor swimming pools soon after 1900 and continued with the introduction of sports halls in the 1950s, followed a uniform pattern everywhere because they were subsidised by central government. At the end of the twentieth century, a new generation of indoor sports facilities was launched, for example the Kardinge complex in Groningen and the Sport Boulevard in Dordrecht.

The newest uniform development also originated in Amsterdam, although following examples in other European capitals. This was the construction of the Arena, a multifunctional (football) stadium as the centre of a service area. All major Dutch cities are currently copying this Amsterdam example as fast as they can. Most of these stadia are public-private or wholly commercial enterprises.

The last two examples demonstrate how an increasing number of sports facilities are becoming detached from the green areas where they originated about a hundred years ago. On the other hand, in many present-day urban parks informal teams play in a way resembling the start of football more than hundred years ago, while parks also host other informal sports activities like jogging and basketball. In contrast, in the Netherlands almost all golf links are situated far outside cities, which implies, just like the manors established centuries earlier, the transfer of park-like scenery from town to countryside.

CHRISTIANE EISENBERG

5 Playing Fields in German Cities, 1900–2000

Sport and green spaces is not a topic which automatically arises in Germany, especially in connection with cities. The thought of a warm summer evening on the cricket green might fill the hearts of Englishmen with sentimental thoughts but for a German it would be atypical to link this with the idea of sport. There are several explanations. First, the absence of a tradition of using common lands as playing fields – because there were no common lands in pre-industrial Germany.[1] Second, the absence of a lively tradition of rural games and pastimes to put its mark on modern sports, as can be observed in eighteenth- and early nineteenth-century Britain. In Britain the children and grandchildren of those immigrants to the growing urban centres, who had once played village football on Shrove Tuesday and continued to cultivate their favourite leisure activity in a new environment, provided the majority not only of the paying public but also of the (professional) sportsmen. In German towns and cities, by contrast, sport was not introduced by immigrants from the countryside, but by British tourists, engineers and students.[2] This cultural transfer from Britain gives us a third explanation for the dissociation of the notion of sport and green spaces in Germany. For the transfer took place between 1890 and 1914, a period marked by heavy industrialisation, rapid urbanisation and the consolidation of a strong national state. This framework of conditions contributed to a shift in the meaning of typically 'English' notions of sport that, among other consequences, found expression in a marked tendency to neglect the 'green' accents of sports spaces.

This chapter will first reconstruct some concrete perceptions and misperceptions in the German sporting world that were significant far beyond the narrow world of competitions in the tense political period between 1890 and 1914. In this connection, some specific differences between German

---

1 On the crucial importance of 'common lands' for the early English sports see L. Chubb, 'The use of common lands as playing fields', *Playing Fields Journal* 1 (1930), no. 2, 34–40.
2 C. Eisenberg, 'German workers and "English sport": some notes on the limits of cultural transfer in the nineteenth and early-twentieth centuries', in I. Blanchard (ed.), *Labour and Leisure in Historical Perspective, Thirteenth to Twentieth Centuries. Papers Presented at Session B-3a of the Eleventh International Economic History Congress, Milan, 12th–17th September 1994* (= special issue no. 116 of *Vierteljahrschrift für Sozial- und Wirtschaftsgeschichte*). Stuttgart: Franz Steiner Verlag 1994, 149–159.

and British ways of looking at sport will be outlined. The study will then examine the predominant features of sports arenas and playing fields in turn-of-the-century Germany. Since the original notion of sport was repeatedly reinforced after 1914, the third part of the essay will describe and explain the metamorphoses of playing fields (and other sporting facilities) during the twentieth century.

The making of a specifically German notion of sport

The export of 'English sports' to Germany at the turn of the century was not an act of British 'cultural imperialism'. Whenever the British abroad needed team players or opponents they tried to mingle with the local population who in turn were often delighted by the opportunity to participate. Since German sporting enthusiasts found it difficult to understand the ideas behind certain rules of behaviour which the British considered as self-evident, they developed their own ideas.[3] The precise meaning of 'fairness', for example, was as alien to them as the gentlemanly virtue of 'disinterestedness'; a virtue which was expressed by the British in playing down any deliberate attempts to improve the performance, and in renouncing the idea of systematic training.

Other shifts in accentuation were a consequence of a different social basis of sport. Like British sport, German sport enjoyed wide support amongst the middle classes while the working classes stayed aloof.[4] However, the German middle classes had developed different ideas about their specific culture. Their way of seeing themselves included certain courtly and military traditions which had either never existed in Britain or died out long before. In Germany, however, they were still extremely vigorous and attached themselves to the new culture of sport. Amongst these traditions were the preference for uniforms and medals, which victorious sportsmen could wear on their jackets as if they were military or social decorations; the 'challenge' to a match which recalled duelling rites; the value placed on the correct ('officers') posture during the riding event; and finally the translation of the British concept of 'fairness' as '*Ritterlichkeit*' or 'chivalry'. In this way sportsmen in Germany bedecked themselves with a particular aura of superiority which could be used as a vehicle to integrate themselves into the established culture and at the same time to subvert it by ironic means.

In addition, German middle-class persons committed themselves to sport in the knowledge that Britain, the first sporting nation, was simultaneously an industrial pioneer and a global power. The ambition of many young Germans was, therefore, not only aimed at taking part in competitions and attaining the standards set by their teachers. They also wanted to beat the

---

3 Where otherwise not stated, the following paragraphs summarise a number of results from my study *"English sports" und deutsche Bürger: Eine Gesellschaftsgeschichte 1800–1939*, Paderborn: Schöningh 1999.

4 The most that can be said is that, in Britain, industrial workers gave football a special flavour at a very early stage in the game's history, whilst in Germany – even in this discipline – the working-classes were conspicuous by their absence; see Eisenberg, *"English sports"*, 209–214.

British at their own game. The term 'record' took on a completely new meaning. Whereas the British understood a 'record' as a certified performance, German athletes took it to mean a hitherto unbeatable performance: and where this new and different quality of sporting activities was accompanied by a preference for cycling and motor sport it bolstered connotations of progress, speed and danger.

German athletes' love of experimenting with the human body was an equally important factor in image-building. This feature, too, was less conspicuous in the mother-country of sport, not only because deliberate physical training was rejected by many gentlemen sportsmen, but also because Britain had no institutionally mediated traditions of rational physical culture. There was no universal military service with the corresponding habituation of broad sections of the population to military drill. Non-military physical activities like gymnastics and *Turnen* which had spread all over Europe and North America since the start of the nineteenth century and were mass movements around 1900, also remained largely unknown in Britain because they had neither been popularised within the state school system nor through a system of clubs which identified themselves with middle-class political movements.[5] The creation of a general consciousness that it was possible to educate, shape and change the body was a significant factor in paving the way for Germans to associate sport with values such as power, stamina and performance.

This particular accentuation was further underlined by the fact that modern natural sciences and applied social sciences were much more developed on the Continent than in Britain during the period when sport was beginning to spread. Whilst students at Oxford and Cambridge Universities who were keen on sport generally tended to follow courses in the humanities (preferably classical languages), a significant number of academic sportsmen in Germany dedicated their studies to modern natural and social sciences. Against this background, sportsmen who were medical students, chemists, clinical psychologists and experts on manpower studies made experiments on themselves in cooperation with the leading researchers of the time. Their interest in bodily experimentation was shared by politicians concerned about public health, and the military establishment who demanded physical fitness. Sports studies therefore developed into a modern academic discipline in Germany even before 1914.[6]

The image of the athlete or sportsman as an *Übermensch* able to call on his own reserves of power and energy was universally popularised, for the triumphal march of sport in Germany occurred simultaneously with the breakthrough of modern mass media. Advertising posters and illustrated

---

5 C. Eisenberg, '"German gymnastics" in Britain, or the failure of cultural transfer', in S. Manz, M. Schulte Beerbühl and J.R. Davis (eds), *Migration and Transfer from Germany to Britain 1660–1914*. Munich: Saur 2007, 131–146.

6 H. Langenfeld, 'Auf dem Wege zur Sportwissenschaft: Mediziner und Leibesübungen im 19. Jahrhundert', *Stadion* 14 (1988), 125–148; J.M. Hoberman, 'The early development of sports medicine in Germany', in J.W. Berryman and R.J. Park (eds), *Sport and Exercise Science: Essays in the History of Sports Medicine*. Urbana: University of Illinois Press 1992, 233–282.

magazines with glossy photographs were soon followed by 'living images' in the cinema, which gave athletes an extraordinary prominence. It does not come as a surprise, therefore, that sport in the young German Reich attracted the attention of the political establishment, particularly as it could hardly be linked with Social Democracy and the labour movement. Appeals from sporting officials for financial support therefore fell on fertile ground with the authorities and many political bodies trying to win over sporting associations to their cause. These bodies included several middle class (and anti-Social Democratic) young people's associations which were financed by the War Ministry.[7]

William II, the German Emperor, also discovered a taste for sport. In the run-up to the 1904 Olympic Games, and again in 1908, he mobilised his aristocratic relations to set up an Olympic Committee, gave substantial financial help to the German team, and encouraged some of the members of his entourage to start building a modern German stadium equipped with the latest technology, including a sporting research laboratory. This stadium, planned to be ready for the 1916 Olympic Games (they were cancelled because of the First World War), was meant to impress the world.

Turn-of-the-century sporting venues

The framework conditions outlined above, led to the fact that in Germany, sport was never regarded merely as a social leisure activity (as it was traditionally viewed in its country of origin), but that as early as 1900 it was raised to the status of a national affair. As can be seen from the example of the 1916 Olympic Stadium, this idea was clearly reflected in the construction of sporting venues. It would, however, be too narrow to emphasise this indirect effect as being simply dependent on this formative period. For the atmosphere behind the rise of sport in Germany – catching up economically, rapid urbanisation, striving for national prestige, the endeavour to make sport a science – also had a direct effect on sporting practices, in so far as it helped to decide on the concrete conditions in which sporting activities took place. This tendency was especially reflected in German sporting venues, most of which – by contrast with Great Britain – were not sited in a 'green' environment.

It goes without saying that the sporting disciplines made known by the British in turn-of-the-century Germany also included a range of disciplines which can generally be subsumed under the idea of 'green sports'. As early as 1900 there was even a special magazine entitled *Der Rasensport. Zeitschrift für Fußball, [Leicht-]Athletik, Hockey, Turnsport, Eissport usw.* (Grass sports. The journal for football, athletics, hockey, gymnastics, ice

---

7 The most important of these bodies were the *Zentralausschuss für Volks- und Jugendspiele* (Central Committee for People's and Young People's Games) attached to the National Liberal Party, and the *Alldeutscher Verband* (Pan-German League) and the *Jungdeutschlandbund* (Young German League), which were financed by the War Ministry and associated with the *Wehrvereine* (militia associations) and *Jugendwehren* (Youth Militia).

Figure 5.1 Hockey match in Berlin, c. late 1920s. (The Carl und Liselott Diem-Institut, Deutsche Sporthochschule Köln).

sports etc.).[8] But anyone reading this magazine today would immediately notice a remarkable gap between theory and practice. Whereas the programmatic articles propagated the benefits to health from being in the open air, and sometimes even promoted a 'back to nature' philosophy, the reports left no doubt in readers' minds that 'grass sports' were very unlikely to bring them much closer to fulfilling this aim. Either sportsmen completely ignored 'grass sports' like hockey (which was only popular amongst the 'higher class of daughters'), golf, (which suffered a similar unsuccessful fate and was only played in international health resorts), and the British national game, cricket (which, much to the regret of the editors of 'grass sports', was of no interest to any of the readers whatsoever); or – and this was equally disappointing to the editors – so-called grass sports were not played on grass at all, but on gravel, clay, ash or asphalt.

The typical venue for football in pre-war Germany – and we must not forget that football was not particularly popular at the time – was a military parade ground.[9] Athletics took place either on cycling tracks or on the streets.[10] Indeed outdoor stadiums which had especially been built for sport

8 Only vol. 4 is still available (in the archives of the German Amateur Athletic Federation in Darmstadt).
9 'Open up the parade grounds! Following a petition from the *Zentralausschuss für Volks- und Jugendspiele* the Prussian War Ministry has issued an order allowing parade grounds in all garrison towns to be used for public sporting activities involving adults and young people, where these activities are regularly practised.' *Jahrbuch für Volks- und Jugendspiele* 3 (1894), 144.
10 See the illustrations in H. Bernett, *Leichtathletik in historischen Bilddokumenten*. Munich: Copress-Verlag 1986.

Figure 5.2 Tennis court in Berlin, mid-1920s. (The Carl und Liselott Diem-Institut, Deutsche Sporthochschule Köln).

only existed in a few cities. And as many of them were erected by businessmen seeking a profit, they were preferably built with grandstands bedded in mortar. In addition, a number of different 'ice palaces' had been built before 1914. These were leisure halls in the middle of big cities, easy to reach by public transport and with a lot of extras like sales kiosks, smokers' lounges and fashion shops.[11] Ice-skating itself took place on an area inside the cycling track which had been sprinkled over with water.[12] Tennis courts were generally built on empty land between buildings which was being kept free for a time from motives of financial speculation. Even where comfortable tennis courts had been erected for an upper middle-class clientele, the game was played on ash courts because grass surfaces did not thrive in Germany as well as they did in Britain. Thus it was not long before the term 'lawn tennis' was replaced by the single word 'tennis'.[13]

Gymnasts too could only dream of grassy playing fields. German *Turnen* was usually exercised in the back of public houses or in indoor gymnasiums. From the start of the nineteenth century when Jahn's '*Turnen* movement' was driven indoors, state schools and grammar schools got instructions from the authorities to equip gymnasiums. The movement was banned from the open air because the authorities wanted to keep *Turnen* out of the public eye due to their liberal and, even at times, socialist ideas. However the authorities

---

11 See illustration no. 19 in Eisenberg, *"English sports"*.
12 F.A. Schmidt, 'Die Spielplätze in Bonn', *Jahrbuch für Volks- und Jugendspiele* 2 (1893), 72.
13 On turn-of-the-century German tennis, see Eisenberg, *"English sports"*, 193–209.

were reluctant to dispense with *Turnen* altogether for they were well aware of the benefits they provided to health, education and military discipline. This was basically why gymnastics halls were erected with public money at such an early period.[14]

The pattern of 'indoors' instead of 'outdoors' and 'grey' instead of 'green' was a prominent feature of German sport for a long time, not least because the halls were put at the disposal of 'apolitical' middle-class clubs either completely free of charge or for a nominal sum. Furthermore they were located in densely populated urban areas, easily accessible by foot, so that no matter what the weather was like they could be visited in the evenings provided they were equipped with sufficient lighting. As early as the first decade of the twentieth century accessibility was an important precondition for the exceptional popularity of *Turnen* and indoor sport in general which attracted more than 1.2 million persons in 1914.[15] In this context demand was successfully pre-shaped by supply.[16]

The German approach to building sporting venues: metamorphoses in the twentieth century

The connection, briefly outlined above, between the framework of conditions for the rise of sport in Germany and the specific features of constructing sporting venues – features that can already be observed in the early phases of the development of sport in Germany between 1890 and 1914 – remained clearly recognisable in the following decades. As the following review of the turbulent history of Germany shows, path dependencies grew up in the area of the construction of sporting venues, whose effects only began to decline around the end of the twentieth century.

1914–1945

The First World War introduced countless soldiers to sport for the first time and led to a boom in the number of participants after 1918, which resulted in an even greater demand for the type of venues known from the pre-war era. This was reflected in sporting statistics which showed that the *Deutsche Turnerschaft* (German Gymnasts' Association), whose clubs basically used the halls, had a total of 1.5 million members at the end of the Weimar Republic (1919–1933), making it the largest single sporting organisation in Germany.

---

14 H. Becker, 'Turnsperre', in P. Roethig (ed.), *Sportwissenschaftliches Lexikon*. Schorndorf: Hofmann 1992 (6th ed.), 538. In a greater detail: Eisenberg, *"English sports"*, 120–123.

15 For the figure, cf. E. Jeran, 'Deutsche Turnerschaft (DT) 1868–1936', in D. Fricke et al. (eds), *Die bürgerlichen Parteien in Deutschland: Handbuch der Geschichte der bürgerlichen Parteien und anderer bürgerlicher Interessenorganisationen vom Vormärz bis zum Jahre 1945*, vol. 1. Leipzig: VEB Bibliographisches Institut Leipzig 1968, 606. For illustrations of gymnastic halls (and other sporting venues in the German Kaiserreich) see T. Schmidt, 'Entstehung und Gestaltungsmerkmale städtischer Sporträume in Berlin von 1860–1914', in W. Ribbe (ed.), *Berlin-Forschungen*, vol. V. Berlin: Colloquium-Verlag 1990, 131–173.

16 This argument developed from contemporary examples is also valid for the nineteenth century, see for example K. Heinemann and M. Schubert, *Der Sportverein: Ergebnisse einer repräsentativen Untersuchung*. Schorndorf: Hofmann 1994, 321 ff.

Turnen enthusiasts were also in the majority in social-democratic and religious organisations. By comparison, the *Deutscher Fußball-Bund* (German Football Association), whose members generally played in the open air, had to take second place with around 900,000 adherents, despite a huge increase in membership.[17]

The First World War did however give rise to one particular innovation, as it caused a shift from privately owned sporting venues to those provided by public authorities. In addition, from the 1920s onwards public money was no longer spent exclusively on building representative facilities like the Olympic Stadium in Berlin or gymnasiums, but for sporting venues in general, indoors and outdoors alike. Given that Germany had lost the war and was forced by the Treaty of Versailles to reduce its army to 100,000 men, many contemporaries considered sport as a welcome replacement for military duties. This explains why city councils increasingly adopted older plans from the pre-war era[18] and built sporting venues on undeveloped land on the edge of towns as well as small, isolated individual projects in densely populated areas.[19] Some major cities like Berlin, Frankfurt, Cologne and Hamburg which had lost their urban defences as a result of the Treaty of Versailles created plans for generous green belts around the inner city areas including the erection of so-called 'sports parks'.[20] This term was chosen to define the concentration of representative stadiums (the so-called 'Kampfbahn' or 'Battle track' – note the military connotations!) and various other sporting venues like swimming stadiums, physical training halls and playing fields, at a central point in the green belt. Here special care was taken to ensure that the whole site made up an architectural ensemble which fitted into the newly-designed landscape.[21]

The 1920s sports parks were conceived along extraordinarily generous lines, the explanation being that they were, to all intents and purposes, part of emergency schemes to get the unemployed back to work. Both Germans and foreigners alike regarded them as exemplary models at the time.[22]

---

17 For the Deutsche Turnerschaft (1929): *Jahrbuch der Leibesübungen*, 36 (1929), 187; for the Deutschen Fußball-Bund (1929): J. Wendt, Grundzüge der Geschichte des Deutschen Fußball-Bundes und des bürgerlichen deutschen Fußballsports im Zeitraum von 1918–1933 (Unpublished PhD thesis, Halle-Wittenberg 1975), 265. See also C. Eisenberg, 'Massensport in der Weimarer Republik: ein statistischer Überblick', *Archiv für Sozialgeschichte* 33 (1993), 137–178.
18 See for example M. Berner, 'Gross-Berlin ein Sportzentrum der Zukunft?', *Illustrierter Sport* no. 6, 10 February 1914.
19 Parade grounds, shooting ranges and garrison courtyards were now state property. Furthermore, following the revolution of 1918, a huge number of estates owned by monarchs and princes were transferred to the property of the fiscal authorities, something that equally benefited the towns. See R. Heyer, *Funktionswandel innerstädtischer grünbestimmter Freiräume in deutschen Großstädten*. Paderborn: Schöningh 1987, 101.
20 For precise details on the growth in areas in individual towns, see Heyer, *Funktionswandel*, 103 f. Cf. also the following case studies: W. Hegemann, *Das steinerne Berlin: Geschichte der größten Mietskaserne der Welt*. Berlin 1930, reprint (abridged) Berlin: Ullstein 1963, 318 ff.
21 T. Schmidt, 'Stadionbauten in Berlin', *Sozial- und Zeitgeschichte des Sports* 4/2 (1990), 75 f.
22 Illustrations in: Deutscher Reichsausschuß für Leibesübungen (ed.), *Deutscher Sportbau*. Berlin: Deutscher Reichsausschuss für Leibesübungen 1930.

Looking at them from the perspective of sport and green space, however, it has to be said that the colour grey was predominant. All in all sporting venues were nothing more than grey patches in an otherwise green environment. Apart from a few grassy fields for ball games and grass-covered football pitches in the middle of the now obligatory multi-purpose stadiums (for football and athletics), sports parks basically consisted of fixed buildings and pitches made from ash or asphalt. Access roads and parking lots also represented incursions into an otherwise green landscape.

With the exception of the generous financial aid (taken from the funds of the dissolved trade or sports unions) pumped into building the Olympic Stadium in 1936, the Third Reich (1933–1945) did not continue with the Weimar Republic's comprehensive programme of building sporting venues activities. That said, following political pressure from the German Workers Front (*Deutsche Arbeitsfront*), the social-political pillar of the Third Reich, a total of 620 factory sports grounds were built by a huge number of companies.[23]

1945–1989

Many of the main sports parks, including the factory sporting venues, were damaged during the Second World War. However it was the simple inner-city sporting venues that suffered most from the bombardments.[24] After 1945 local authorities gave priority to rebuilding these as quickly as possible.[25] In this respect the tried and trusted indoor sites once again took on a high level of significance. Alongside conventional physical training halls, there were now for the first time large-scale multi-purpose sporting venues for handball, indoor football and other ball games. Their numbers increased by around 300 per cent between 1960 and 1988.[26] Starting in 1960 the venues were erected on the basis of a 'Golden Plan for Health, Sport and Recreation' which had been worked out by the *German Olympic Association*, and for which regional and local authorities had set aside extra funds. Around 17 billion DM was invested in sporting venues up to 1975,

---

23 T. Mason, *Sozialpolitik im Dritten Reich: Arbeiterklasse und Volksgemeinschaft*. Opladen: Westdeutscher Verlag 1978 (2nd ed.), 252. For the general trend see J. Eulering, 'Staatliche Sportpolitik – aus der Sicht der Länder', in H. Ueberhorst (ed.), *Geschichte der Leibesübungen*, vol. 3/2. Berlin: Bartels & Wernitz 1981, 872.

24 For precise data see G. Breuer, *Sportstättenbedarf und Sportstättenbau: Eine Betrachtung der Entwicklung in Deutschland (West) von 1945 bis 1990 anhand der baufachlichen Planung, öffentlichen Verwaltung und Sportorganisation*. Köln: sb 67 Verlagsgesellschaft mbH 1997, 27–29, 77–83, 120–127.

25 See for example B. Almstedt, 'Beispielgebend im Sportstättenbau: Sportförderung der Stadt Hannover seit 1945', in Niedersächsisches Institut für Sportgeschichte, Hoya e.V. (ed.), *Sport in Hannover von der Stadtgründung bis heute*. Göttingen: Die Werkstatt 1991, 202–211.

26 Breuer, *Sportstättenbedarf* 127. Unfortunately the author fails to provide any overview on the development of sporting venues in absolute figures. Furthermore he complains that, for the time after 1990, there are no useful data available (108). – Because of the differing methods of calculation, the great number of figures on the number of areas available per inhabitant are of no use; cf. C. Diem, 'Spielplatz um die Ecke: Der Zauber einer Statistik/ Sport und Spiel des deutschen Volkes (1938)', in C. Diem, *Ausgewählte Schriften*, Bd. 2. Köln: Academia Verlag 1982, 241–242.

Figure 5.3 Turnhalle Zehlendorf, a modern multi-sport gymnasium, c. 1960s. (Landesarchiv Berlin).

and a further three billion DM between 1975 and 1980. Some local authorities even provided more funds than they were officially obliged to do.[27]

The basic difference between sporting venues financed with funds from the Golden Plan and the older buildings was that the new ones were principally built to meet the needs of competitive sport. Germany had been split into two parts after the Second World War, and during the Cold War the conflict was more acrimonious on the running track than anywhere else.[28] As a result, the sporting venues met the standards for sport of *DIN* (*Deutsche Industrie-Norm*, German industrial standards). This feature was even more noticeable because the experts who had put forward the Golden Plan for the sporting venues had themselves thought out the norms with great zeal; one of the tasks of the *Bundesinstitut für Sportwissenschaft*, founded in 1970, was to support these efforts [29]

As a consequence, the venues constructed in Germany during this boom period were not only thoroughly functional but they also looked more or less the same. The planners were not primarily interested in green spaces and decorative details. At best they experimented with coloured stripes on the buildings and with plastic cladding, a characteristic feature especially popular in the 1970s. In this way the magnificent sporting venues of the

---

27 Cf. Eulering, 'Staatliche Sportpolitik', 873.
28 See U.A. Balbier, *Kalter Krieg auf der Aschenbahn: Der deutsch-deutsche Sport 1950–1972. Eine politische Geschichte*. Paderborn: Schöningh 2007.
29 Cf. H.-H. Kämmerer, 'Sportstättenbau und Sportgeräte – Bestandsaufnahme, Entwicklungsperspektiven', in Bundesinstitut für Sportwissenschaft (ed.), *20 Jahre Bundesinstitut für Sportwissenschaft 1970–1990. Symposium am 26. September 1990 in Köln*. Köln: Sport und Buch Strauß 1991, 70.

Figure 5.4 Suburban sports facilities, c. early 1970s. (The Carl und Liselott Diem-Institut, Deutsche Sporthochschule Köln).

period were no longer grey but colourful – yet not 'green' in the natural sense. Against this background the sole sports park project of the period, the Olympic Park in Munich with its daring tent-like buildings, seems to be doubly spectacular.[30]

1990–2000
It was not only after the end of the Cold War in 1989 that the sporting venues built in the 1960s and 70s came to be generally regarded as

---

30 The planners foresaw that the Munich Olympics Park should be a part of the urban landscape. It was to be integrated into the neighbourhood as part of a major green belt beginning three kilometres north of the Central Station and running through to the man-made lakes in the north of Munich. Parts of this project were later realised: W. Jerney, 'Münchner Parks seit Olympia', *Garten und Landschaft* no. 9, 2002, 13.

unimaginative ('shoe boxes') and full of deficiencies, if not to say out-of-date.[31] Certainly the new political situation arising from German reunification helped to foster this perception. Even the keenest supporters of more recent Golden Plans (like the one for the reconstruction of East Germany) have long conceded that sporting venues have to look different nowadays.[32] However, apart from the fact that plans to meet modern demands could not be implemented anyway because public bodies have been forced to make large budgetary cuts as a result of the general economic crisis, it is questionable whether the construction of lavish sporting venues is in harmony with the spirit of the times. For many decades now there has been a decline, not only in the population as a whole but also in organised sporting activities in general, a process that at first went unnoticed because young people were drawn into sport in the post-war years, and later, in the 1960s, as a result of a growth in the number of women participants. Nowadays there is a clear tendency towards informal, individualised sports and anyone interested in sport has a choice of around 240 different disciplines, many of them esoteric, to say the least![33] Some people have simply abandoned their clubs and associations – the traditional beneficiaries of public venues – in favour of private fitness studios and other commercial organisations. Others practice their sporting activities in the streets, on pavements and in parks.

From the perspective of sport and green space, this last aspect – informal and individualised sporting activities in parks and urban wooded areas – is the most noticeable development in the area of urban sport. In Germany it is one of the by-products of the trend towards less formal behaviour which started in the late 1960s where 'No Trespassing' notices were largely ignored and the population as a whole began to re-possess green spaces in a quiet revolution. Now the planners' traditional distinction between 'decorative' and 'healthy green spaces' has become obsolete – as has the role of 'green protectors', a product of the German idea of organisation and order in this area.[34] Thus sporting activities in Germany have finally become 'green' after more than a century.

The development has in no way been free of conflict, as evidenced by growing criticism that the growth in informal sporting is detrimental to the

---

31 It is claimed that 40–50 per cent of sports grounds are desperately in need of repair; cf. J. Eulering, 'Grundzüge des Handlungskonzepts Sporträume in NRW', in Die Grünen im Landtag NRW (ed.), *Sportstätten und Agenda 21: Dokumentation des sportpolitischen Fachgesprächs vom 21. Juli 2004*, no place, no year, 14.
32 Eulering, 'Grundzüge', 15, and J. Eulering, 'Sportstätten – Zur Entwicklung der Bewegungsumwelt', in K. Heinemann and H. Becker (eds), *Die Zukunft des Sports: Materialien zum Kongreß "Menschen im Sport 2000"*. Schorndorf: Hofmann 1986, 147–172. See also Heinemann and Schubert, *Sportverein*, 321 ff.
33 Cf. Heinemann and Schubert, *Sportverein*, 168. Specially on landscape intensive leisure sports, see W. Strasdas, *Auswirkungen neuer Freizeittrends auf die Umwelt: Forschungsbericht der Technischen Universität Berlin, Institut für Landschafts- und Freiraumplanung. Im Auftrag des Bundesumweltministeriums.* Aachen: Meyer & Meyer 1994.
34 These "green protectors" were mostly older men, often invalids, who took care to ensure that nobody ran over the grass, and even prevented children from doing so. For nineteenth century park concepts, see: D. Hennebo, 'Der Stadtpark', in L. Grote (ed.), *Die deutsche Stadt im industriellen Zeitalter*. München: Prestel 1974, 77–90.

environment, primarily because green areas are being used for a huge range of different activities. Whilst athletes consider they are only leaving footprints and informal tracks behind them, others regard this as damaging 'nature'. And where sporting participants shout encouragement to each other, others long for nothing more than peace and the melodic sound of bird-songs. The criticism of sport as being harmful to the environment has given rise to dozens of official enquiries and provoked legal judgements in the areas of noise protection and statutes in favour of nature protection.[35] This should also be seen against the background that towns have been able to acquire very few new open spaces since the second half of the twentieth century.[36] Municipal parks are becoming increasingly crowded and the different users are beginning to tread on each other's toes.

Conclusion

The new trend also explains why golf courses have been shooting up all over the place in the last few years. As we shall see in Chapter Eleven, golf in Germany has traditionally been underrepresented but it now comprises the most interesting new development in the area of sport and green spaces – albeit exclusively outside town centres. On the other hand, golf and other 'green' sporting activities merely indicate that a minority of Germans interested in sport have turned their backs on the 'grey' tradition of German sporting venues. The majority are still forced to indulge in sporting activities on urban sites in the middle of residential areas (that will be primarily built from stone and concrete in the future), and will remain grateful for the opportunity to do so. We should therefore be wary of placing too little value on the tradition of building sporting venues. Compared with other countries – not least Great Britain, where modern sport originated – this tradition has made a significant contribution to the fact that contemporary Germans are more actively involved in sport than are the residents of other European nations.[37] Official membership figures issued by the German Olympic Sports

---

35 See for example W. Nohl and U. Richter, *Umweltwirkungen durch vermehrte Freizeiteinrichtungen in der Stadtentwicklung: Im Auftrag des Umweltbundesamtes*. Berlin: Umweltbundesamt 1992; Strasdas, *Auswirkungen*, especially 57 f. For legal measures see Eulering, 'Sportstätten', 165, and R. Knauber, 'Gemeinwohlbelange des Naturschutzes und Gemeinwohlgebrauch der Landschaft durch Sport: eine rechtliche Würdigung', in Arbeitsgemeinschaft beruflicher und ehrenamtlicher Naturschutz e.v. (ed.), *Sport und Naturschutz im Konflikt* (= Jahrbuch für Naturschutz und Landschaftspflege 38, 1986), 155–175.
36 Some towns have enlarged their leisure areas through mergers with smaller towns. The resulting improved orientation values are however only of statistic importance, given the unchanged state of affairs in densely populated urban residential areas; cf. Heyer, *Funktionswandel*, 133.
37 I. Hartmann-Tews, *Sport für alle!? Strukturwandel europäischer Sportsysteme im Vergleich: Bundesrepublik Deutschland, Frankreich, Großbritannien*. Schorndorf: Karl Hofmann 1996; C. Eisenberg and T. Mason, 'Sport und Sportpolitik', in H. Kastendiek and R. Sturm (eds), *Länderbericht Großbritannien: Geschichte – Politik – Wirtschaft – Gesellschaft – Kultur*. Bonn: Bundeszentrale für politische Bildung 2006 (3rd, revised ed.), 391–408.

Association (*Deutscher Olympischer Sportbund*) in May 2006, list 90,000 clubs and 27 million individuals. But even if we reduce these figures somewhat (because they may be exaggerated), it is extremely impressive that 33 per cent of the population is involved in organised sport.[38] Arguably, whether sporting venues are 'green' or 'grey' provides no help in determining the recreational and health values involved in sporting activities.

---

38 Cf. http://www.dosb.de/fileadmin/fm-dosb/downloads/bestandserhebung/DOSB_ Bestandserhebung_2006.pdf (last accessed 25.1.2008).

FULVIA GRANDIZIO

6 Green Space and Sport in Italian Cities in the Twentieth Century: the Example of Turin

In Turin, green spaces came to be considered a public good from the beginning of the nineteenth century. Once there was the walled city, and green space was private and small-scale, the privilege of convents and the palaces of the aristocracy.[1] Green areas available to everybody were outside the walls, but they were thought of as 'the country' and were not linked to the idea of leisure, but to work and physical effort. Futhermore, these areas were often insecure, or even dangerous places. Outside the city, a short distance from the capital, stood the summer estates of the House of Savoy, including Venaria Reale, Stupinigi, Racconigi, and Agliè. These residences, which had been built from the end of the sixteenth century, were well-known as the *Corona di delitiae* (crown of delights), because they were laid out around the city. These castles and palaces surrounded by large parks were created as the hunting preserves of the dukes and the leisure places of an inner elite.[2]

Between the end of the eighteenth century and the Congress of Vienna, the French ruled the city: they pulled down the baroque walls and replaced them with great scenic avenues. Turin was opened up to the outside world and to the River Po, which for centuries had almost been ignored. The city looked beyond the river, to the Hills, which until then had been the exclusive holiday resort where the aristocracy built its mansions. With the walls torn down, the city had a huge amount of ground on which to expand, but it had neither the funds, nor a pressing need to do so.

In general, for most of the nineteenth and twentieth centuries, green space, despite the positive official rhetoric, was a low priority for city authorities and decision makers: too often it was just the urban space left over when all the city's building and commercial needs had been met. In this scenario outdoor sports spaces were slow to develop, except where they catered for

---

1 On Turin see A. Cardoza and G. Symcox, *A History of Turin*. Turin: Einaudi 2006; M.D. Pollak, *Turin, 1564–1680: Urban Design, Military Culture, and the Creation of the Absolutist Capital*. Chicago: University of Chicago Press, 1991; V. Comoli Mandracci, *Torino*. Bari-Roma: Laterza 1983; A. Cavallari Murat, *Forma urbana ed architettura nella Torino barocca dalle premesse classiche alle conclusioni neoclassiche*. Turin: UTET, 1968.
2 C. Roggero Bardelli et alia, *Ville Sabaude*. Milan: Rusconi 1990.

the activities of the elite. In this chapter we look first at early moves to develop parks and green spaces; then at the advent of new sports and their impact on policies towards green space; this is followed by a discussion of green space and sport in the Fascist era, and the role of the Fiat company in promoting sports and recreation areas; and finally the last three sections consider green space and sport in the post-war era, along with the effects of urban growth and planning; the growth of green policies from the 1970s, and the recent expansion of golf courses.

Early initiatives

Although the first public park projects date back to the first half of the nineteenth century, only a few became a reality as they were expensive and it was considered better to sub-divide the land for construction. Turin grew south-eastward, towards the Po. Here the 'enlargement' plans started, and in 1835 the first public park for the 'people's pleasures' was born: the Garden of the Ripari.[3] This green space was located on a section of fortifications where demolition costs would have discouraged builders. A second landscaped park, the 400,000 square metre Valentino Park, was laid out by French architect Pierre Barillet-Deschamps along the left bank in the second half of the nineteenth century. It became the favourite place where high society met and the lower classes took their leisure.

Except for the gardens in squares, urban green planning was carried on until the twentieth century according to the rule that only land not suitable for building should be given over for green spaces.[4] After years of debate, the new urban planning regulations for the city and its environs were passed in 1913, and the green space issue was finally included in a Royal Decree of 15th January 1920. The planners' aim was to create a system of six large parks that would be 'shared so that each area had a park, and that these areas are ideally suited to become parks as they are not suitable for building on and their cost is low'. Four green areas would be located along the old city customs barrier, while two parks would be laid out along the rivers. Not every park became a reality, but in the long drawn out process of approving the planning regulations, many debates and conflicts led to important changes, and at the same time exposed the different interests at stake and the sensibilities of the actors.[5] Some members of the City Council sought to show that from the 1880s to the 1910s the city had taken no important initiatives in the field of public green spaces. As far as the large public parks were concerned,

---

3 F. Panzini, *Per i piaceri del popolo: L'evoluzione del giardino pubblico in Europa dalle origini al XX secolo.* Bologna: Zanichelli 1993.
4 P.L. Ghisleni and M. Maffioli, *Il verde nella città di Torino.* Turin: Stamperia Artistica Nazionale 1971.
5 Lingotto Park (now, Millefonti-Italia '61 Park, 365,000 sq.m.) was set out on the left bank of the Po; it is the ideal extension of Valentino Park. The Pilonetto Park (136,000 sq.m.) was placed on the right bank, in front of Lingotto Park. The Pellerina Park with an area of 930,000 sq.m. was spread along the Dora Riparia River. The San Paolo district hosted a green area near the railway, today called Ruffini Park (315,000 sq.m.).

private citizens were worried about the expropriation of their land. They hoped that once the city decided which lands to purchase, it would buy them up in the quickest time possible. They thought it would be wrong to stop a landowner using his land only to have it expropriated thirty years later at the old price.

However, in Turin there were also citizens with philanthropic concerns. At the beginning of the twentieth century, a private citizens' committee submitted to the city authorities a project to transform the space of the old parade ground into a public park for the working and lower-middle classes. The proposal was ignored. Between 1913 and 1930, the largest part of the parade ground was broken up and high-class houses with gardens were built. When the International Exposition was held in Turin in 1911, the residual area became the site of the Stadium: a huge sports complex (361 metres long and 204 wide), with two semicircles, surrounded by terraces. It was a multipurpose facility built in reinforced concrete serving as a hippodrome, an athletics track, a football ground, an area for large parades, and a 200 metre swimming pool: a huge building which once the exposition was over was no longer used and soon started to deteriorate. The ruins were demolished, and after the Second World War the area became the site of the Polytechnic.[6]

Sport without green spaces?

The first idea of sport in Turin was born when King Carlo Alberto needed to improve the soldiers' performance and training in the Savoy army. Young men from the lower classes had to put up with extreme working conditions (as many as sixteen hours a day) and their poor health was also due to dietary and hygiene deficiencies. So the king, who held that the best preparation began with a soldier's body and his nervous equilibrium, introduced gymnastics. In 1833 Carlo Alberto called to his court Rudolf Obermann, a famous Swiss instructor, to direct the artillery Military Gymnastics School. Obermann's methods mainly consisted of figurative floor exercises to be conducted in groups indoors. 'The concept of competition was not considered at all, nor was the idea of practising sports in the open air: two trends regarded as dangerous individualising tendencies in Protestant England, to be viewed with suspicion and firmly opposed'.[7] At that time policies in the Kingdom of Sardinia were still conservative, and it was thought English sporting ideas would corrupt the citizen soldier that Carlo Alberto wanted to train up.

A direct consequence of Obermann's presence in Turin was the foundation of the first Italian (and European) gymnastic society in 1844: the *Reale Società Ginnastica*.[8] The gymnastics project was meant for the upper and upper-middle classes, that is, for those with free time to practice sport and

---

6 A. Sistri, 'Spazi, luoghi, architetture', in ASCT, *Torino e lo sport: Storie luoghi immagini*. Turin: ASCT 2005, 289–309.
7 M. Crosetti, 'La città che inventò lo sport', in ASCT, *Torino e lo sport*, 147–173.
8 O. Clerici, 'La Società Ginnastica di Torino', *Torino* 4 (1930), 280–286.

the means to buy club shares. In 1844, on the site that some years later would be known as Valentino Park, the *Reale Società Ginnastica* built the first gymnasium and rented a practise ground. The Society was seen in Italy and abroad as an example to follow, and in 1878 a state law made physical exercise in schools compulsory.[9] Using its membership income the club also organised other sporting activities: in 1846 a target shooting school and swimming lessons in the Po and, in 1848, a fencing school. Turin became the Italian city where the earliest sports groups in many disciplines were founded. In 1897 the *Reale Società Ginnastica* created a new open-air gymnasium for fencing lessons in the Citadel gardens. A debate over the Prussian or the English concept of sport began. Angelo Mosso, a physiologist who maintained the need and the benefit of physical exercise for women as well as men, criticised the Prussian military-type indoor sport and supported the English concept. He suggested the idea of sport for the masses. In his opinion English sports played in the open air – football, cricket, rugby, lawn tennis and rowing – would serve 'not only as an escape valve for youthful physical exuberance [...], but also provide the nation with model citizens, accustomed to discipline, to decorum and also [...] to a spirit of initiative'.[10]

At this time, ordinary people spent the little free time they had, after long working days, playing *boccie* (Italian bowls), a very popular sport in Piedmont played in small areas outside inns. In the second half of the nineteenth century the game became popular among the upper classes and several *boccie* areas were laid out in Valentino Park. Once again the park became the main sports centre in Turin. It was also chosen as the main site for the events at the Exhibitions in 1884, 1898, 1902 and 1911.[11] From the 1870s Valentino Park hosted the *Circolo Pattinatori Valentino* (Valentino ice-skating society). The municipality leased the artificial lake to the skating club until 1933, when the water was replaced by a jumping track for the local riding club. In 1882 the *Veloce Club Torinese* (the Turin Speed Club), a forerunner of cycle racing in Italy, also made its home in the park.

The first rowing society in Turin was founded in 1863. Its aim was to transform boat trips upon the River Po from a leisure activity into a physical exercise. The club was called *Cerea*, from the Piedmontese dialect greeting used between boats during river trips. In a few years, the number of rowing associations multiplied. The best-known clubs were the *Caprera*, *Esperia* and *Armida*, which were later joined by *Medora*, *Eridano*, *Diana* and the *Società Canottieri Ginnastica*. Though all were private, they were located on land leased from the municipality. Between 1969 and 1971 the municipality proposed to evict them, arguing the clubs had to move out because the inhabitants of the district near the river called for more public green and

---

9 On sport see: P. Ferrara, 'Ginnastica, sport e tempo libero', in U. Levra (ed.), *Storia di Torino. VII Da capitale politica a capitale industriale: 1864–1915*. Turin: Einaudi 2001, 1067–1085; S. Pivato, 'Sport', in *Guida all'Italia contemporanea: 1861–1997. IV Comportamenti sociali e culturali*. Milano: Garzanti 1998.
10 Crosetti, 'La città che inventò lo sport', 153.
11 C. Roggero Bardelli, 'Dal giardino privato al parco pubblico nella città', in F. Bonamico et alia, *I giardini a Torino: Dalle residenze sabaude ai parchi e giardini del '900*. Turin: Lindau 1991, 35–48.

leisure spaces. Underlying this dispute there was a long-standing class conflict. Even though club officers spoke about their clubs' democratic composition, which was vouched for by the variety of members' social origins, the membership fees were too high for workers, and the upper middle class pre-dominated in these associations. In 1910 the prefect of Turin noted that the members of the *Esperia* club were: 'All distinguished people of the professional sector, industrial employers and tradesmen, belonging to the party of order'; *Caprera* members were almost all 'rich artisans, engineers, doctors and lawyers', while the members of *Cerea* were forbidden to 'bring people not of good social standing' into the club or take them out on boats.[12] Up to the 1970s there was little change. Even the dispute over land use in 1969–71 was settled by a further renewal of the city's leases to the clubs. It was said that the rowing clubs had to stay beside the river banks and that there were no alternative places fit for the purpose. However, the other private sports associations along the river including tennis clubs, swimming pools and *boccie* grounds, could have been moved away from the river, ensuring they no longer obstructed views of environmental significance. But the public authorities were unable or unwilling to go against private interests, especially those of some of the oldest and the most prestigious and elitist associations in the city.

Green areas in the Fascist era

In the early twentieth century the urban environment was the main cause of the high mortality rate in the city. Sport was seen as important not only for recreation and socialising but as an instrument of public health. In Turin since the second half of the nineteenth century, Don Bosco and his Salesian oratories began a Catholic concept of recreational and physical activities. At the same time, even though often hostile to sport as such, the working-class movement produced its first attempts at independent sporting organisations.

Between 1922 and 1942 Mussolini's regime took over these ideas and made sport and open-air life one of the strong points of its politics. In order to bring up a race of men who were physically and morally healthy the Fascist Party pursued policies to improve infant and maternal health. The regime carried out projects to create some of the parks that had been included in the urban plans made in the beginning of the century. Parks were to be for public use and an open-air lifestyle. They often hosted a *colonia elioterapica* (sunray treatment holiday camp), a place where children and young people could go out in the sun to fortify their bodies and, with physical exercise, protect themselves from tuberculosis, the most widespread illness of the

---

12 A. Cardoza, *Aristocrats in Bourgeois Italy: The Piedmont Nobility, 1861–1930*. Cambridge: Cambridge University Press 1997; A. Cardoza, 'Tra casta e classe: clubs maschili dell'élite torinese, 1840–1914', *Quaderni storici* 2 (1991), 363–388; G. Pécout, 'Les Sociétés de tir dans l'Italie unifiée de la seconde moitié du XIXe siècle', *Mélanges de l'Ecole française de Rome. Italie et Méditerranée 102–2* (1990), 533–676.

time. Natural botanic species were preferred to show that the municipality took care of public expenditure and to display national pride. The Gerolamo Napoleone Park (today Parco Ruffini), a 155,000 square metre green space in the San Paolo district, was planned during the early twenties. From 1937 to the post-war years it contained a *colonia elioterapica* with a swimming pool and playing fields. Work on the Pellerina Park started in 1934 but it was still unfinished in the late sixties. The Millefonti Park was laid out from 1937 with the high-sounding name of Bosco dell'Impero (Empire Wood).[13] In the Hills, the Villa Gualino Park started life as the private estate of industrial baron Riccardo Gualino. After his bankruptcy it remained unfinished. In 1936 it was bought by the Fascist association *Federazione dei Fasci di combattimento di Torino* and transformed into the *Colonia elioterapica 3 gennaio*.[14] The Rimembranza Park, another hillside park, was quickly planned and completed in 1928, as it was intended to commemorate (ten years after the peace) the soldiers who fell during the First World War. In the fascist period, the goal to transform the Hills into a web of parks gradually took shape with the purchase of historical parks of private mansions such as Villa Genero, Parco Leopardi, Villa Rey and Villa Abegg.[15] In the mid-1930s the regime's green spaces programme produced over 100 kilometres of tree lines. To celebrate the twentieth anniversary of the regime the aim was to have trees in every city avenue. Unfortunately, many of these trees would be used as firewood during the Second World War.

The fascist regime saw sport, in addition to the provision of green spaces, as the most appropriate activity for the 'physical and moral improvement of the race'. Competitiveness in sport and combat were put on the same footing by the regime. Practising sport was not only a question of mere physical effort but it was also a moral exercise. '[…] willpower, a life of sacrifice and being used to risks, all made the sportsman healthier, better-trained for toil and danger': the best virtues for a citizen and a soldier.[16] The regime established national regulations for sports bodies. The *Comitato Olimpionico Nazionale Italiano* (*CONI* – the Italian National Olympic Committee) was declared a fascist body, the *Sports Doctors Federation* and the *Chronometers Federation* were founded, and sports insurance was made compulsory for all athletes. Under Fascism, various sports activities, which, until then, had

---

13 E. Grammatica, 'Torino, città ringiovanita dalle opere fasciste: i parchi e i giardini', in A. de Rosa (ed.), *20 anni di fascismo in Piemonte*. Turin: Edizioni di Orsa 1939, 22–27.
14 Two camps were opened in the nearby mountain valleys: the first for 70 children, the second for 800. In those years the main industrial concerns, including FIAT, and other public and private bodies also built camps in the mountains in the Province of Turin and at seaside resorts. In 1935, 286 children enjoyed themselves at the permanent camps. This figure rose to 716 in 1936 and then to 978 in 1938. In the same period, the Piedmontese summer camps hosted 11,148, 14,229 and 18,025 children, respectively. By 1939, there were 58 sunray treatment day camps. M. Escard, 'Il Dopolavoro', in de Rosa (ed.), *20 anni di fascismo in Piemonte*, 47–49.
15 M. Maffioli and L. Re, 'Il giardino del '900, il giardino nel '900: parchi e giardini nella città moderna', in Bonamico et alia, *I giardini a Torino*, 67–80.
16 R. Manganiello, 'L'attività sportiva', in G. Di Giacomo (ed.), *Panorami di realizzazioni del fascismo*. Rome: Casa Editrice dei Panorami di realizzazioni del fascism 1942, Vol. VII: L'azione di governo e le attività nazionali, 429–436.

Figure 6.1 The FIAT mountain camp at Sauze d'Oulx. *(Italian Archivio e Centro Storico Fiat).*

only been practised in Italy as elite sports or at amateur levels, were organised into federations following directives in the 'Carta dello Sport' (Sports Charter) – a political document which left very little room for freedom or sports autonomy.

In the 1920s and 1930s, public and private initiatives enriched Turin with a number of sports centres. As elsewhere in Italy, football and cycling were the most popular sports and both were practised in state-of-the-art facilities.[17] The Velodrome was built in 1920 alongside the Po to a design by architect Vittorio Ballatore di Rosana, a specialist in sports complexes who was well known for his Stadium for the 1911 Exposition. The open-air Velodrome had a concrete track, the maintenance costs of which were high and difficult to meet, and the track was later abandoned. The Velodrome is a listed building and the sports field inside the track is now used for American football matches.

17 On football see: J. Foot, *Calcio: A History of Italian Football*. London: Fourth Estate 2006.

Two football grounds which were to make history went up at around the same time. The 'Campo Torino', also known as 'Filadelfia', was inaugurated in October 1926. Count Marone Cinzano, president of the Torino team, commissioned Ballatore di Rosana with the design. The ground where the team both trained and played became legendary during the 1940s, years when Torino won five championships in a row. Torino's fame ebbed following the 1949 air disaster when the whole team died. Only training sessions were held in the ground from 1963 until the early 1990s, when the team's management decided it was no longer useable, and the ground was partly demolished in 1997.

In 1933 Mussolini decided that the Littorial Games – to celebrate Year XI of Fascism – were to be held in Turin.[18] An idea to use the old Stadium was cast aside and Turin was forced to build another more appropriate venue for the Games in record time. Work began in late September 1932 and, just eight months later on the 14th of May, right on time for the start of the Games, the new venue was opened. The new sports complex, planned by the Turin City Department of Works, included the 70,000 capacity Mussolini stadium, the Marathon Tower, an athletics track, two basketball courts, an indoor and three outdoor swimming pools.[19] After the war the stadium changed its name and became the 'Comunale' Stadium; here *Juventus* held its home games until the early 1960s when *Torino* also began playing their matches there. In 1990, the Football World Cup matches needed a larger stadium and so the 'Comunale' was given up. However, thanks to the 2006 Winter Olympics, the 'Comunale', now called the 'Olympic Stadium', was fully renovated and used for the Opening and Closing Ceremonies of the Games. Today it is once again being used as a football ground by *Torino* and *Juventus*.

In conclusion, though both green space and sport received greater public attention under the Fascists, there was only limited overlap. Most of the new sports areas were enclosed arenas – constructed in concrete grey – rather than outdoor green spaces.

FIAT: organising employee sports and recreation activities.

FIAT was founded in Turin in July 1899.[20] The first factory was founded by a group of some of the leading members of the aristocracy and professional and financial classes. Giovanni Agnelli, a wealthy landowner from near Turin, became an investor almost by chance, but by 1902 he had become the managing director of FIAT.[21] The company soon realised that sports events, which attracted an ever increasing number of spectators, were the best way to advertise its products. The first step was to take part in the numerous car races. To promote the production of bicycles the company

---

18 The Littorial Games, which had been held from the 1930s onwards, were culture, sports and labour events at which university students participated.
19 M. Ceragioli, 'Il Civico Stadio Mussolini', *Torino* 7 (1933), 4–23.
20 Fabbrica Italiana Automobili di Torino (Italian Automobile Factory Turin)
21 V. Castronovo, *FIAT 1899–1999: Un secolo di storia italiana.* Milan: Rizzoli 1999

assembled a group of good amateur cyclists who competed in and won various competitions on FIAT bicycles. In 1911 FIAT enrolled its own team of professional cyclists in the Giro d'Italia race. In the period leading up to the Great War, FIAT also sponsored athletes, one of whom became the Italian marathon champion in 1914.[22]

Factory life changed after the First World War. Workers grew restless at the spread of Fordist work practices, and feared that the scientific organisation of labour would force them to continue with the exhausting shifts and frozen wages they had suffered during the war. The period which became known as the 'two red years' started with strikes in 1919 and reached a peak in 1920 with factory sit-ins, aimed at bringing about the hegemony of the working class on the model of the Russian soviets. Union disputes and revolts, with exchanges of gunshots leaving people dead and wounded, ended in 1921. All the FIAT factories, including the huge Lingotto factory,[23] which was still being built, were involved in workers' struggles and revolts. In addition to fighting for wage increases, the workers also sought guarantees concerning working hours (the agreement on an eight hour working day dated back to 1919), sickness and injury benefits, and pensions – in other words, quality of life.

The *International Labour Organisation* (*ILO*) stated that to improve a worker's quality of life, his free time had to be organised, and he had to be protected from alcoholism, tuberculosis, venereal disease and gambling. Because at first no government initiatives were forthcoming, the responsibility was assumed by associations which sprang up from the *Catholic Action movement.* In Turin in 1919 an autonomous group of FIAT workers set up *USOF* (*Unione Sportiva Operai FIAT* – The FIAT Workers Sports Union) and its members took part in walking races and cycle competitions which they organised with the support and encouragement of FIAT and Giovanni Agnelli Senior. Although the Fascist regime, which had come to power in 1922, considered sport as of the utmost importance for the affirmation of racial qualities, FIAT saw sport more as part of the company's heritage, as a way of improving the welfare and cohesion of the work-force and of stopping worker unrest. This policy was first put into practice when the land that FIAT owned alongside the Po in Corso Moncalieri was transformed into an area for company leisure activities. The 1785 square metre area, the former Diatto railway works, which FIAT had purchased in 1918, was given to workers to use for their free time to play *boccie* and football. A few years later in 1923, Agnelli inaugurated the FIAT Rowing Section in one of the buildings which was equipped as a gymnasium, with meeting rooms and a boat store. In addition, in 1919 Agnelli gave his employees a sports field located in Corso Stupinigi, on the southernmost outskirts of the city, which was first used by a local football team.

This was the basis for the informal founding of the *FIAT Sports Group* in the summer of 1920. The association was formally established by FIAT

---

22 G. Paparella, *1919–1999: Ottant'anni di sport e tempo libero alla FIAT*. Turin: Ce.d.A.S FIAT 1999.
23 Opened by King Victor Emanuel III in May 1923 and visited by Mussolini in October.

Figure 6.2 The Dopolavoro built by FIAT, Corso Moncalieri, 1929. (Italian Archivio e Centro Storico Fiat).

employees in 1922 after the unions accepted the idea that workers, above all the young, could and should practice sport; the idea that was spread and propagated by the *Sports International* founded in Moscow in 1921. Even though company–worker relations were tense in that period, initiatives in the sports-recreation sector led to significant developments in welfare. The *FIAT Clerks' Association* was founded with welfare, education and sports recreation aims in June 1920. Six months later the members founded the voluntary *FIAT Clerks Mutual Aid Society* which was granted legal recognition in February 1921 and was to receive financial support over the years from the company. The *FIAT Labourers Mutual Aid Society*, created in autumn 1923 also received support from the company. Around the same time, the *Agricultural Cooperative* was founded and FIAT handed over land on the outskirts of Turin free of charge. This land was used to grow vegetables which the members either consumed themselves or sold to other members at fixed prices.

The Fascist Party created the *Opera Nazionale Dopolavoro – OND* (National After Work Institute) on the 1st of May in 1925 in order to oversee at a national level all state and company recreational activities.[24] After a

---

24 According to the regime press the 'Dopolavoro' was an 'institution of the Regime, for the physical, cultural and professional improvement of the working classes'. In the arts sector the 'Dopolavoro' organised amateur dramatics groups, musical bands and choirs, set up film shows and conferences for propaganda on the Italian Empire, on autarchy and the programmes of the regime. A number of company clubs organised language lessons, night schools and adult literacy courses. One important aspect was the setting up of vegetable gardens, or allotments, which led to a large amount of hitherto abandoned land being cultivated. By 1939 in Turin and its Province 2,000 families had an allotment which supplied their vegetables. These vegetable gardens covered 300,000 sq.m. See S. Gatto, 'Il Dopolavoro', in de Rosa, *20 anni di fascismo in Piemonte*, 41–46.

Il nuovo grandioso Dopolavoro aziendale che sorgerà alla Fiat-Mirafiori.

Figure 6.3 The Mirafiori Dopolavoro project, 1937. *(Italian Archivio e Centro Storico Fiat).*

'wait and see' policy, to safeguard the autonomy of its own institutions as long as possible, FIAT brought together all the FIAT sports groups under a single institution called *Dopolavoro FIAT* in 1928. The company held out until 1934 when the *Dopolavoro* officially became part of the *OND*. As the numbers of participants grew the leisure area along the Po soon became too small. In 1929 FIAT sports activities moved to a new 12,000 square metre centre, designed for 600 members, which had been built quite close to the original centre along the Po.

In 1932 FIAT also built a sports hall in one of its old warehouses in Via Marocchetti, the site of the first FIAT factory. Four years later the centre in Corso Stupinigi was renovated and a new athletics track was opened. The Giovanni Agnelli Trophy was established in 1931 for the factory which obtained the best employee sports results. After the Mussolini Stadium was inaugurated in 1933, Agnelli decided to compete with the regime by commissioning (in 1937) plans for a huge 275,000 square metre sports complex in the Mirafiori area next to the new factories. A thirty metre viewing platform was also planned for the *Mirafiori Dopolavoro* and the complex would have included a park with a children's area, a lake and a track for children's scooters and bicycles. However, the Mirafiori sports complex was never built, and the plans were abandoned after the war when the Mirafiori factory was enlarged.

At the end of the war in 1945, during the transition period before the Allies arrived in Turin, the *Comitato di Liberazione Nazionale – CLN* (Italian National Liberation Committee) entrusted FIAT management to four commissioners. The *Dopolavoro FIAT* became the independent *FIAT Sports Group*. The Regional Council established the *Mutua Azienda Lavoratori FIAT – MALF* (FIAT Employees Mutual Aid Society) which brought together the workers' and the clerks' mutual aid societies. After a few name and role changes and management changeovers between 1948 and 1952, the *Centro Sportivo Ricreativo e Culturale FIAT – CSRC* (the FIAT Sports, Recreation and Cultural Centre) was founded on the 1st of January in 1953. The company was once more free to manage the organisation of its members' free time. The number of employees rose from 70,000 in 1952 to 150,000 in 1972, while *CSRC* membership rose from 14,204 in 1953 to 69,237 in 1967.

Figure 6.4 The FIAT swimming pool centre, Corso Moncalieri. (Italian Archivio e Centro Storico Fiat).

In 1955 the athletics field at the Corso Stupinigi sports centre was renovated to conform to international competition standards. A sports hall was built on the Via Guala side in 1969. FIAT built a new sports centre in the Borgo Vittoria district on the outskirts in 1953 and added a swimming pool and gymnasium in 1972. In 1969 the 50-year lease from the city of Turin of some of the buildings along the Po came up for renewal, and as the cost of the new lease was too high, FIAT handed over all the buildings to the city. In 1973 FIAT built two sports centres at numbers 336–346 Corso Moncalieri, including four swimming pools. FIAT also built residential complexes with sports facilities in the suburbs where they had some of their factories.[25] The *CSRC* was split into two parts in 1976, the Sports Centre and the Social Activities Centre. The former became *SISPORT FIAT* in 1978 and was tasked with managing sports competitions and the ownership of the sports centre. The latter was in charge of organising recreational sports and cultural activities for FIAT employees and their families. Both centres are still running.

Green areas and sport in Post-Second World War Turin

Between 1951 and 1971 Turin's population rose from 721,000 to 1,164,000. Migration into urban areas led to the decline of living conditions for city dwellers. During the late 1960s, letters of protest to newspapers, the claims of district councils, applications to the Sports Department to rent grounds, appeals to the municipality by technicians and experts from every political alignment created a critical imbalance between the supply and demand for sporting facilities within the city. Some of the main demands included: 'To redress the ecological balance of the urban environment, to create new green spaces, to provide parks, gardens, playgrounds for the rising neighbourhoods, to defend the surviving green area …'.

The City Council realised that the city must provide spaces that allowed access to nature and contained areas for playing sport. Outdoor sports activities were increasingly considered not merely as competitive pursuits but as a means to develop the human personality from a psychological and physical perspective. In 1971 the Public Works Department carried out a preliminary survey for a plan to improve the urban environment from the viewpoint of sports and green space availability. For the first time in its history the city took a snapshot of the facilities of both sectors. The current urban planning regulations for public green space were studied and the state of all public and private sports activities was registered.[26] The survey highlighted deficiencies linked to two causes: legal delays in handling the provision of public green spaces and sports facilities, and the slow development of public opinion on the two issues. The inadequacy of legislative measures can be outlined as follows. The 1150/1942 planning law, which included the drafting of general town planning regulations, ignored public green and sports topics. In 1957, Law 1295 created the Institute for Sporting Credit and highlighted the need to bring sport to the masses in a non-competitive and non-spectator way. In 1968 a Ministerial Decree set a 9 square metre per inhabitant quota for green and sport (together) – falling to 4.5 square metres in war damaged areas.

Under Fascism, Law 383/1934 dealt with public gardens' upkeep but made no mention of the creation of new parks and sporting facilities. They

---

25 There were few public sports centres in the chief cities and virtually none in the provinces. Every morning during the school year, the FIAT sports facilities were available to school children (even non FIAT employee children) who lived in those cities.
26 Assessorato LL. PP. Comune di Torino, *Il verde e lo sport a Torino*. Turin: Comune di Torino 1972.

were considered an optional expense and were not permitted in cities with a budget deficit. The town planning regulations implemented in 1959 set the ratio for green spaces and sport at 10 square metres per inhabitant (2 square metres for local green spaces without sport, 1 for local sport, 3 for urban green spaces, 2 for urban sport, and 2 for special and private sporting facilities). In 1964 a Ministerial Decree updated the cadastral rent ratio. Sporting facilities increased to the value of 60, being on a par with a luxury home. Finally in 1972 a Presidential Decree handed over the management of sporting facilities of regional interest to the Regions.

Only in 1955 was a Sports Department instituted by the city. The survey made in the 1970s showed that the amount of urban green space was inadequate. There were only a few free spaces inside the city: the green areas of the inner city districts could not be improved, and the suburban areas would have to be used to create large leisure complexes. Furthermore, the statistics regarding sports activities were greatly below expectations. Given the existing number of grounds and facilities, less than 11 per cent of city dwellers played a sport in a meaningful way for 2–3 hours a week across all sporting facilities (municipal, private, school and military). This was less than half the ideal figure of 20–25 per cent that experts called for.

The situation was particularly acute since heavy urban growth necessitated the construction of a huge rash of Council and private housing for workers in the suburbs. This forced urbanisation meant swathes of agricultural land were used up, while in the centre and densely built-up areas green space was an indefeasible right for the inhabitants and disputes to establish ownership became struggles between public and private bodies.[27]

Studies for conurbation planning began in the late-1950s in order to deal with the problems connected to the city of Turin and its surrounding areas. Particular attention was placed on the environment and landscape. It was planned to provide 6,000 hectares of parks with existing and new agricultural areas, woods, natural parks, play and leisure areas, river banks and the hillside as the linking elements throughout the area. Unfortunately, the plan was not approved.[28]

Meanwhile, Turin was getting ready for the centenary celebrations for the unification of Italy, particularly important since the city had been the first capital of Italy in 1861.[29] The event was given the name *Italia '61*, with plans to show the progress the country had made since its defeat in war just sixteen years earlier. The site chosen was the long neglected Millefonti Park along the river opposite the hills, which had become a sort of shantytown

27 In Turin and the Hills, the growth of the green area came about through the acquisition of the parks of ancient aristocratic mansions. More significant examples are the park of seventeenth-century Villa Rignon, bought by the city in 1955, and Villa 'La Tesoriera', which became publicly owned in 1976. The latter was in the middle of an 18-year dispute involving the typical castes of the times: on one side the ruling Christian Democrat Party, the Savoy family and the Jesuits and, on the other, the opposition left-wing parties, with protesting students in the background.
28 A. De Magistris, 'L'Urbanistica della grande trasformazione (1945–1980)', in N. Tranfaglia (ed.), *Storia di Torino: IX Gli anni della repubblica*. Turin: Einaudi 1999.
29 S. Pace, C. Chiorino and M. Rosso, *Italia 61: The nation on show. The personalities and legends heralding the centenary of the Unification of Italy*. Turin: Allemandi 2005.

Figure 6.5 View from the hill of Italia '61 *Exposition (Historical Archives of the City of Turin).*

where many immigrants lived. Despite resistance, the bulldozers knocked down the shacks and the inhabitants were given housing in the suburbs. The exhibition was held in pavilions designed by some of the most famous architects and engineers of the time.[30] However, a few years after the celebrations, the exhibition area was abandoned and it was not until the late-1970s that some of the buildings were given new uses, and others were demolished, though the rest waited years before being given a new lease of life and usage.

From the 1970s: the swing towards the environmentally sustainable city

After the Second World War, industrial growth accelerated alongside the city's waterways, which became dumps for toxic waste. Fish and birds disappeared. The rivers were dying, but in the 1970s and 1980s they were saved by the setting up of a purification plant. Road infrastructures modified the few surviving natural features of the river banks. The only link between rivers and city-dwellers were the immigrant workers' illegal vegetable patches. A reversal of the trend began between the late-70s and the early-80s, when the new Regional Urban Law and the first studies for the new

---

30 The most important exhibition buildings, which are still standing, were named Palazzo del Lavoro (Labour Building), designed by Pier Luigi Nervi, and Palazzo a Vela (Sail Building). A Japanese-like monorail train wound its way through the exhibition area and a cable car linked Italia '61 to a park in the Hills.

Urban Plan of Turin treated nature and the environment with a new concern. Architects, town planners, municipal officers, dweller planning committees and environmentalist associations started the first studies to create a network of rivers and hillside parks. An Urban Plan was introduced in 1993 and was soon followed by the *Torino Città d'Acque* (Turin City of Water) project.

The *Torino Città d'Acque* plan has a wide ranging urban and regional influence. At the urban level, the plan is to create a continuous system of river parks stretching 70 kilometres along the urban banks of Turin's rivers (the Po, Dora Riparia, Stura di Lanzo and the Sangone). The parks are to be interlinked by cycle tracks, pedestrian and educational paths. Another aim is to create a transition belt between the urban parks and the more 'natural' environment of the hilly and suburban parks. The *Torino Città d'Acque* project is linked with that of the project *Corona Verde* dealing with regional parks.[31] In 2000 the Piedmont Region approved this programme of safeguarding natural landscape and cultural heritage. The aim is to create connections within the city and between the city and its hinterland through an ecological corridor linking historical places and inner cities, royal estates, rural architecture and wildlife reserves. The plan involves some 24,000 hectares of nature reserves (16 areas), while the architectural heritage consists of all the urban and suburban House of Savoy residences. The latter are the major point of the project and include the recently reopened Venaria Palace, the biggest European investment in renovation, funded by the Region, the Italian Ministry of Cultural Heritage and Environmental Conservation, and the European Union.[32]

However, the many river parks to be gradually completed by the City of Waters project will have no sporting facilities. The bulk of them are abandoned areas and the primary aim is to restore the biodiversity they had before post-war pollution. Inner city parks along the river Po are places where people can go jogging, but they are unsafe. In some areas it is common to see drug peddling. Since the 1971 survey, there has been no significant increase in the number of public outdoor sporting facilities. Over the last ten years, however, there has been a boom in private health clubs where people can do indoor sports such as circuit training and aerobics. These gym clubs are seen in an individualistic way, as places where personal health care coincides with aesthetic fulfilment. Those who have a predilection for outdoor sports and contact with nature travel every week-end to the nearby mountains where they can go rock-climbing, trekking, canoeing, mountain biking and, in winter, skiing; or go to the country for golf and horse riding in the several golf clubs and riding schools that have arisen in recent decades.

---

31 A. De Rossi and G. Durbiano, *Torino: 1980–2011. La trasformazione e le sue immagini*. Turin: Allemandi 2006.
32 Regione Piemonte – Settore Pianificazione Parchi, Comune di Torino – Settore Verde Pubblico, 'Corona Verde–Torino Città d'Acque', *Acer* 6 (2001).

The spread of golf courses

Golf made its official debut in Italy when *Rome Golf Club* was founded in 1903. The *Menaggio* and *Cadenabbia golf clubs* on Lake Como were founded in 1907. The *Italian Golf Federation* (FIG) was founded in Milan in 1927 as a private association by the *Florence, Palermo, Stresa, Torino* and *Villa d'Este clubs*. By 1920 there were 20 clubs and the federation moved to Rome where it became part of *CONI*. The Italian Federation became a member of the *European Golf Association* in 1936. In 1954 there were 17 clubs with 1,200 players. Ten years later there were 31 clubs which had risen to 90 in 1988 and by 1999 there were 258 clubs. At the close of 2006 there were 84,117 members in 241 clubs.[33]

The location of golf courses in Italy reflects the distribution of wealth in the various regions. Thus most courses and players are in the large prosperous regions of northern Italy in Lombardy, Piedmont, Emilia Romagna and the Veneto. Sardinia is the exception in the southern regions and islands and has twelve courses, linked to the development of exclusive tourism and the involvement of foreign investors over the past few decades.

Golf in Italy has always been a sport for the limited few and played in exclusive clubs. The last twenty years has seen an increased interest in golf mainly by young members of the upper-middle and upper classes, who enjoy lifestyles allowing them travel opportunities for work and pleasure. The increase of golf has given rise to two opposing schools of thought, the first represented by golf players and investors, and the second by environmentalists. According to golfers, a new golf course set in a beautiful landscape close to cultural areas will attract tourists from overseas, while a golf course in an area of no particular environmental value is mainly for local users. Environmentalists are often radically opposed to golf and point out the close ties with property development. They also say that the construction and running costs of a golf club are so high that to be economically viable nearby land has to be used to build apartments and villas. Ecologists criticise the public funds golf courses receive via the backdoor through the way Regions subsidise investments in small-medium hotel and tourism businesses in depressed areas. Again there are complaints that golf courses are not subject to normal planning procedures to evaluate the impact on the environment. Golf courses are often created in or near areas of natural value which, according to environmentalists, are put at risk by the intensive use of pesticides. Irrigation of golf courses is an issue which particularly effects the south of Italy where there is a chronic lack of water and poor management of water resources. Often when a golf course is in a tourist area along the coast wells will be drilled to water the course automatically which causes seepage of salt water into the water-bearing stratum, and so damages the local ecosystem and farming activity.[34]

---

33 http://www.federgolf.it.
34 See A. Atzori, 'Campi da golf? No grazie!' in http://www.project-s.it/golf-atzori.htm, updated 2004 and the website http://www.antigolf.org

In order to promote golf in Italy and catch up with English speaking and northern European countries and also with Spain, Portugal and North Africa, the *FIG* has taken steps to promote an eco-sustainable image of golf. In 2000, Italian golf adhered to the 'Committed to Green' environmental project and took its first steps to come into line with the directives of the *European Golf Association* to protect the environment and support the ecological management of golf courses.[35] Since then the Green-Keeping section of the National Golf School started a course of theoretical-practical lessons to train personnel how to work in Italian weather conditions which are not favourable for grass growing. It is up to the clubs themselves whether to participate in the Committed to Green programme. The clubs undertake to carry out improvements by following a management plan established by an environmental consultant taking account of the course features.[36] Certification is granted by a national technical committee and the European advisory commission. Clubs which receive recognition are subject to periodic checks to make sure they maintain standards. A number of clubs have enrolled in the project since 2000. The *Carimate Golf Club*, in the Province of Como, and the *Verona Golf Club* received national certification for eco-compatibility in 2001 and the following year they received European certification.

Aiming to verify the ecological value of golf courses and their role in nature protection, in 2001 twenty-three Italian golf clubs carried out the first survey of bird species found on golf courses. In some cases, the survey also included surrounding areas to make a comparative evaluation. The case studies selected were in diverse environments: twelve were agricultural, seven urban and four were coastal areas. When a comparison was made of the wildlife on a golf course with that in the surrounding areas with similar natural conditions (such as in coastal areas) the results showed the golf courses coming out on top; with protected species, results were substantially similar. Results from golf courses in urban and agricultural settings showed there were more birds on the golf courses than in the adjacent areas as they offered more variety than the urban areas or the areas of monoculture. Fifty per cent of the species were on the high priority preservation list: this was helped by hunting being prohibited and by courses having limited public access. Hence golfers believe golf courses play a similar role to city parks. However if golf courses have an important role in wildlife protection and if, as such, they are preferable to city parks, they lack one essential democratic condition: they do not allow free access to everyone and they are places of elitist segregation.

The region of Piedmont has 49 golf courses which take advantage of the variety and beauty of the regional landscape. Seven clubs are at high altitudes in the mountains. The most well known of these is the *Sestrieres Golf Club*, founded in 1932, under the auspices of FIAT, in the Susa Valley at 2035

---

35 S. Verde, 'Una convivenza ben riuscita', *Acer* 3 (2007), 30–35; P. Caggiati, S. Di Pasquale, V. Gallerani, D. Viaggi and G. Zanni, *Gli effetti ambientali delle attività ricreative sul territorio: Il caso del golf in Italia*. Bologna: Ge.S.T.A. & D.E.I.AGRA 1999.

36 These points are: protection of nature, the landscape and cultural heritage; water resources management; green keeping; refuse management; energy saving; training for staff and the environment; public awareness.

above-sea-level. Nineteen clubs are in the hills. Some are set among the vineyards of the provinces of Asti and Alessandria, while others are in the woods and hills in northern Piedmont. Three of these are in panoramic positions near the lake district. There are 14 courses in the countryside (plus one with a pitch and putt and driving range). The *Stresa Golf Club Alpino* – founded in 1924 – has a flat course in the country-side near Lake Maggiore. The *Verbania Golf and Sporting Club* lies on the shores of Lake Mergozzo. Three courses are close to Turin and in medium sized towns such as Asti and Alba with a further five courses in the suburbs of Turin and other chief cities of Piedmontese provinces. The *Torino* and the *Roveri Golf Clubs* are in the former Venaria Royal Hunting Lodge grounds, now the Mandria Regional Park. The *Roveri Golf Club* (now renamed *Royal Park Golf & Country Club*) considered one of the best courses in Italy, has 18 holes designed by the leading American golf architect Robert Trent Jones to which a further 18-hole course, designed by Michael Hurzdan Fry, has been added. The *Torino Golf Club*, founded in 1924, is the oldest Turin golf club and had three homes before finally moving to the Mandria Park. Close by to the park there is also a nine-hole course and two driving ranges, one of which is on a noble's former agricultural estate which served as a sort of home farm for the Royal Reserve. Throughout the region there are a further six clubs which were created on land formerly belonging to, and in some cases still belonging to, the aristocracy. One of these is the *Castel Conturbia Golf Club*, in the Province of Novara, which was the first private golf course in Piedmont. This club was founded by Count Gaspar Voli in 1898 as the Couturbier Golf Course. After having visited Scotland a number of times, he passed on his passion for golf to his neighbour Count Avogadro di Collobiano, who created the first nine-hole course in the grounds of his seventeenth-century castle. Members of the Royal House of Savoy used to play on the course when they met at Conturbia for foxhunting. The course was much appreciated for the excellent way it exploited the landscape.[37]

At present golf in Piedmont is enjoying a period of growth and, as elsewhere in Italy, this had led to a conflict of interests. As far as urban golf courses are concerned, Turin is taking a pioneering stance. In September 2007 the municipality of Turin and the *Italian Golf Federation* (*FIG*) started a partnership to improve the 700,000 square metre Stura Park. This riverside park, which is still being laid out, had earned the dubious name of Toxic Park as it is the meeting place for addicts and pushers. The city of Turin and *FIG* are working to create a 9-hole course on 29 hectares of the park. This course will be run jointly so that the course remains an integral part of the park. Whether it will be a club or a 'pay and play' course has yet to be decided. This second option would make the course less elitist and more accessible to the less affluent, such as those who live in the nearby districts. As such it would be the exception to prove the rule that golf courses in the Turin area and elsewhere in Italy are largely a private preserve of the urban affluent classes.

---

37 http://www.piemontegolf.it

Conclusion

As the story of golf underlines, the development of public green space in general, and outdoor sports spaces in particular, has been a slow and difficult one in Italian cities like Turin. With the exception of a small number of municipal parks, most open green spaces and outdoor sports facilities have been in private hands, often controlled by the elite and commercial interests. Under the Fascists there was some attempt for ideological and health reasons to expand the public provision of green space and recreational areas, but in Turin the major increase of provision of outdoor sports facilities sprang from the paternalist initiative of the FIAT and other smaller industrial companies, concerned to improve the welfare and productivity of their workers. After the Second World War despite new efforts at planning in the city and region, urbanisation ran well ahead of any increase of green space and recreation areas, and the participation in sport by city inhabitants remained low. Only after the 1970s did the situation improve with the growing influence of environmental groups and the European Union. In the Turin area a new network of river parks improved both biodiversity and informal public recreation, but the provision of public sports facilities within the city remains a relatively low concern. Rather those interested in sport are expected to be wealthy enough to join private health clubs or leave the city and travel to the mountains or countryside to go mountaineering, riding or to play golf.

Acknowledgements

I am grateful to Giuseppe Paparella for his help and the FIAT Historical Archive for kindly supplying illustrations. Many thanks to Paolo Odone for his precious information and to Walter Gaino – Library of the Department of Agronomy, Forest and Land Management of Faculty of Agronomy of Turin. Special thanks to Filippo De Pieri for his patient review of my paper and his helpful comments. Thanks to Andrew Martin Garvey for editorial assistance with this paper.

PART II:
Sports Areas, Planning and Environment

JÜRGEN H. BREUSTE

7 Green Space, Planning and Ecology in German Cities in the Late Twentieth Century

Germany is one of the European countries that have a long tradition of urban landscaping. The destruction of many German cities during the Second World War, and the country's political division by the Iron Curtain played an important role in post-war urban development, and the planning of urban green spaces during the second half of the twentieth century. The extended wastelands caused by destruction during the war contributed to the development of scientific studies on urban natural conditions, urban ecosystems and their management. This was especially the case in the isolated city of West Berlin, cut off by Communism from West Germany. In the 1970s, the natural science discipline of urban ecology was developed by the ecologist Herbert Sukopp and his school in Berlin, influencing the field of ecology first in Europe and then worldwide. In the view of the ecologists, management and planning of urban ecosystems (biotopes, green spaces and so on.) should be undertaken on the basis of the comprehensive spatial mapping and registration of characteristics of all nature elements in the city (biotope mapping). This led to an increase in the amount of attention given to urban nature conditions in Western Europe in the mid-1970s and fostered nature protection and nature management in urban areas in many European cities[1]. Berlin, the birth place of urban ecology and capital of Germany, now has more than 2,500 public green and open spaces totalling about 5,500 hectares embedded in forested natural surroundings (Regional Parks), and can be seen as an example of a Green City.

Urban landscape can be defined as the landscape of urban settlements and their surroundings, characterised by urban land-use forms, often called urban and peri-urban or metropolitan landscapes. In Germany and many other Central European countries, urban expansion and sprawl means these

---

1 L. Trepl, 'Flora und Vegetation', in H. Sukopp et al., *Ökologisches Gutachten Rehberge Berlin*. Berlin (West): 198452-69 (= *Landschaftsentwicklung und Umweltforschung* 23); L. Kowarik, 'Das Besondere der städtischen Flora und Vegetation: Natur in der Stadt - der Beitrag der Landespflege zur Stadtentwicklung', *Schriftenreihe des Deutschen Rates für Landespflege* 61 (1992) 33–47; J. Breuste, H. Feldmann, O. Uhlmann (eds), *Urban Ecology*. Berlin, Heidelberg: Springer 1998, 714; J. Breuste, 'Urban ecology', in O. Bastian and U. Steinhardt (eds), *Development and Perspectives of Landscape Ecology*. Kluwer Academic Publishers: Dordrecht 2002, 405–414.

landscapes occupy increasing amounts of space. Furthermore, in an unstoppable process, a growing number of people are becoming urban dwellers. In Germany, urban landscapes have expanded into former agricultural and forest landscapes in the vicinity of cities, creating a patchwork of urban, quasi-rural and rural spaces. Urban landscapes are no longer limited to the administrative borders of existing towns and cities, but include a 10 or more kilometre wide transition zone surrounding many cities or bigger towns – the peri-urban zone.[2] Figure 7.1 shows the distribution of urban landscapes in Germany, where the total growth rate of urban land use (built-up areas and transportation lines) averages 70.4 hectares per day.[3] Modern urban landscapes thus consist of a mixture of land-use forms: specifically urban land-use such as residential or industrial areas, agricultural land-use, and forest landscapes, as well as other types of land-use.[4]

One driving force of contemporary urban growth in Germany is economic power and the concentration of infrastructure and services. Although this process is not new, it has at the present time enforced a more dynamic and intense land-use than 50 years ago, supported by the widespread use of automobiles.[5] This allows the location of residential areas at an increasing distance from workplaces, often in more natural environments near forests, lakes and rivers. Meanwhile, as the city grows, green spaces, often the only nature in a highly urbanised landscape, come under growing urban pressure from increased densities and expansion.[6] Yet green spaces continue to play an important role in improving the urban environment and serving growing recreational needs. During the last four decades these functions of urban green spaces have been increasingly recognised by urban administrators, been investigated by urban ecologists, and are now considered an essential element of a general concept for urban well-being and health. Germany is one of the European countries which paid early scientific and practical attention to green spaces as a multi-functional urban amenity. In consequence, the development and management of urban green spaces, their location,

---

2   J. Breuste, 'Landschaftsschutz – ein Leitbild in urbanen Landschaften', in H.-R. Bork, G. Heinritz and R. Wießner (eds), *50. Deutscher Geographentag Potsdam 1995*, vol. 1, 1996, 134–143; J. Breuste, 'Indicators of urban ecology', in T. Deelstra and D. Boyd, (eds), *Indicators for Sustainable Urban Development. Delft 1998* (= Proceedings of the Advanced Study Course on Indicators for Sustainable Urban Development, 5th–12th July 1997, Delft, the Netherlands), 1998, 79–88.
3   J. Breuste, 'Nutzung als Untersuchungsgegenstand und Raumbezug der Stadtökologie', *Natur und Landschaft* 33(2/3) (2000), 95–100.
4   J. Breuste, 'Die Stadtlandschaft - Wandel und Perspektive einer Kulturlandschaft', in Bayerische Akademie für Naturschutz und Landschaftspflege ANL (ed.), *Laufener Seminarbeiträge 4/95* (= *Vision Landschaft 2020: Von der historischen Kulturlandschaft zur Landschaft von morgen*), 1995, 63–74; Breuste, 'Indicators of urban ecology'.
5   M. Hesse and S. Schmitz, 'Stadtentwicklung im Zeichen von "Auflösung" und Nachhaltigkeit', *Informationen zur Raumentwicklung* 7/8 1998, 435–453; H. Mäding, 'Entwicklungsperspektiven für die Stadt – Trends und Chancen', *Difu-aktuelle Information*, Dec. 1997, 1–11; T. Sieverts, 'Die Stadt in der Zweiten Moderne, eine europäische Perspektive', *Informationen zur Raumentwicklung* 7/8 1998, 455–473; T. Sieverts, *Zwischenstadt: zwischen Ort und Welt, Raum und Zeit, Stadt und Land*. 2. ed. Braunschweig, Wiesbaden: 1998.
6   Breuste, 'Landschaftsschutz'.

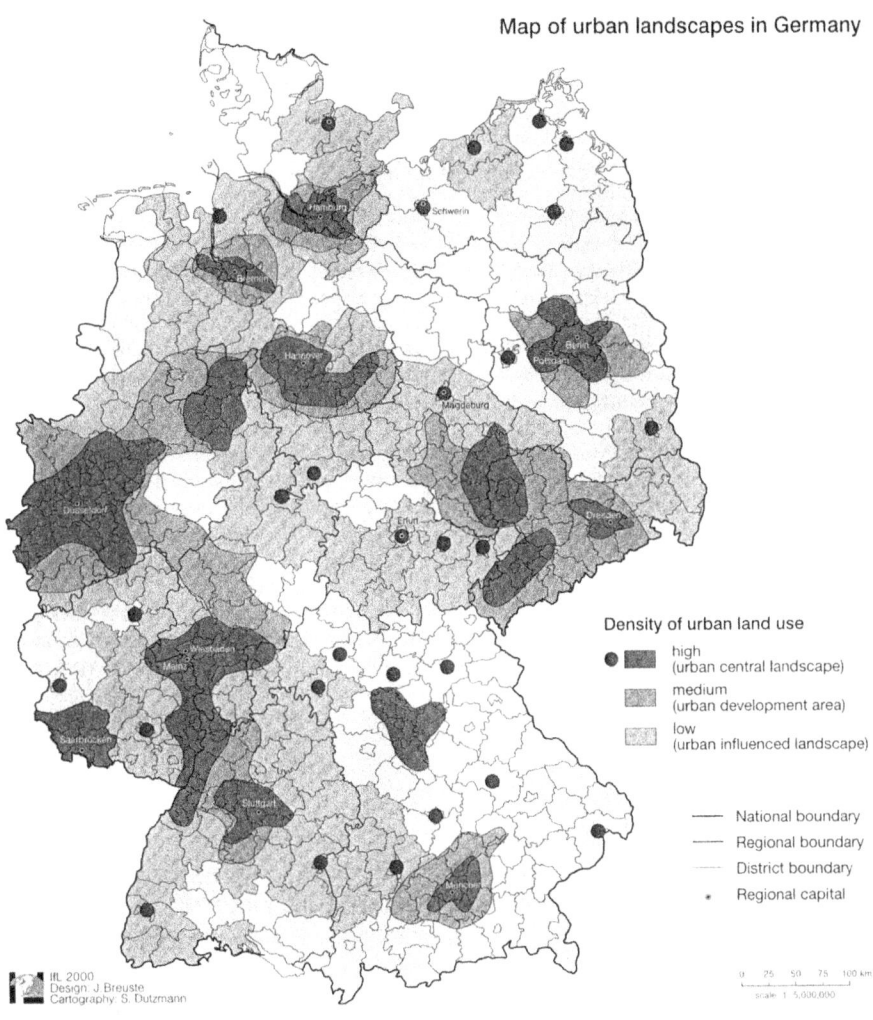

Figure 7.1 Urban Landscapes in Germany. Source: J. Breuste, 'Nutzung als Untersuchungsgegenstand und Raumbezug der Stadtökologie', Natur und Landschaft 33 (2/3) (2000), 95–100.

structure and use is a well established planning and management objective. Often this goal is in conflict or tension with the commercial forces of urban development. While municipalities may not always handle this issue in an effective way, contemporary German society widely supports the development of urban green spaces. However, to understand how all this came about we must look first at the problems and failures of urban green space development and planning after 1945, and then at the new trends in urban nature conservation and environment education in the 1970s, and their impact in the last decades of the twentieth century.

General trends and planning concepts of green space development after the Second World War

After the war, the planning concept for green spaces was largely focused on extending green areas to keep pace with general urban expansion, and correcting imbalances between urban districts. These objectives were rarely achieved because of inadequate financial support for municipal parks and garden departments. At times, imbalances in the distribution of green spaces between the older, densely built-up areas (former working-class districts) and new residential areas became acute in German cities. Meantime, city centres often displayed grand, commemorative green spaces to show their prosperity and prestige (especially in East Germany) or to celebrate their restoration and renewal (particularly in West Germany).

In East Germany, the political rhetoric of green space went further, with the emphasis on sport as a major means of demonstrating the power of the political system. A large number of outdoor sports grounds were erected in the newly built up outer areas of East German cities. Sports grounds were effectively planned zones to which the political decision makers paid special attention. Every new residential area was equipped with extensive sports grounds, in accordance with the population number of each district. Older scenic parks were also sometimes redesigned to include sports facilities, as shown in Figure 7.2.

Flood plains were in many cases chosen as the location for new and extended parks in East Germany. These flood plain areas were not previously used intensively, because of periodic flooding. Yet despite the extension of green spaces for sports and as part of the residential expansion in the 1960s, by the 1970s one sees a clear reduction of green spaces in East Germany because of a chronic lack of funding. Investment in urban green management could not keep pace with what was needed for the expansion of built-up areas. Many spaces designated for development as green spaces were left for decades as vacant and blighted land in the midst of residential areas. Only at the beginning of the 1980s did the concept of interior urban green development for densely built-up urban districts receive more attention, influenced by developments in West Germany (for example, the greening of inner courtyards).

After the reunification of East and West Germany in 1990 sports like golf, which up to this time had been discouraged for political reasons, could now develop in the surroundings of many former East German cities. The impact of commercialisation and social differentiation in urban sporting activities is more and more evident (see Chapters Five and Eleven by Eisenberg and Eisenberg and Tamme).

In West Germany as early as the later 1940s urban planning regulations had been implemented in the federal states, which gave a recognised role to urban green spaces. All federal states of West Germany created institutions for the planning and management of urban green spaces. The planning of urban green spaces as an integral part of city Master Plans was generally achieved in the 1950s. The most widely accepted concept was that of 'green networks', which is comparable to a road infrastructure network. This showed

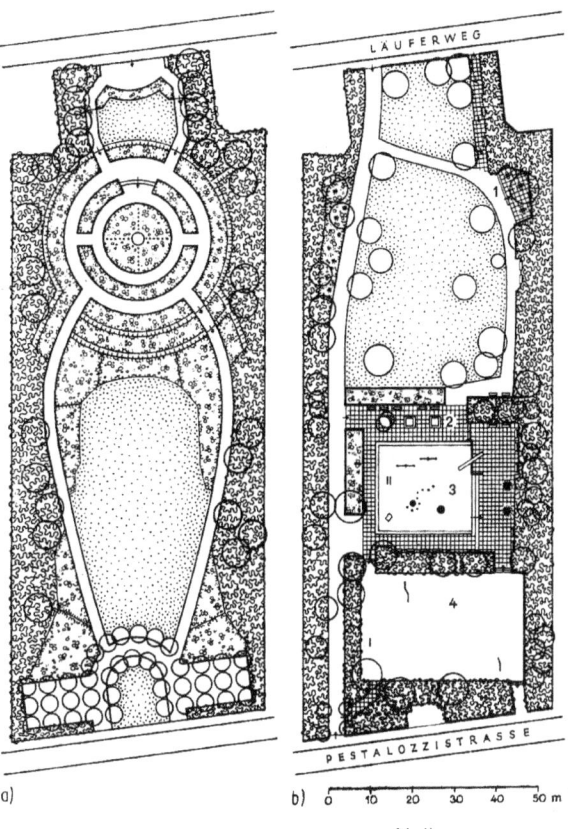

Figure 7.2 Location of outdoor sports in urban parks – the case of the remodelled Pestalozzi-Park in Halle/Saale (Germany)
a) Previous state, Draft: Mengel
b) New installation: 1. Skating ground; 2. Sandbox; 3. Sand ground with playground equipment (slide, climbing frame, climbing tower, leapfrog, see-saws, bars, elephant); 4 Ball playground, Draft: Därr, Ziegler
Source: J. Greiner and A. Gelbrich, Grünflächen der Stadt. Berlin: Verlag für Bauwesen 1997, 132, fig. 142

the high degree of adoption of urban green space planning already in the 1950s and 1960s. Green space was seen as an aesthetic and important design element organising urban space into a number of smaller-scale intertwined built-up areas which were separated by green corridors, including networks of streams, natural relief features, parks and sports grounds. These corridors also connected the city with the surrounding landscape. This widespread concept influenced city planning in West Berlin until the start of the 1970s.

However, economic development pressures became too powerful at the end of the 1960s, and the 'green network' idea was abandoned in most West German cities and replaced with a supposedly more realistic planning strategy, leaving green space planning under the shadow of other urban planning priorities. A phase of removal or restructuring of open spaces and green corridors followed at this time. Vacant land, agricultural land, allotments and natural landscapes, were frequently used as new building locations.

Green spaces as focal points of urban nature conservation and environmental education

By the mid-1970s, the economic-driven destruction of nature (not only in cities) was no longer acceptable by West German society. From the end of the 1960s there had been mounting radical and resident protests at urban redevelopment policies (as in other European countries) and an embryonic green movement gathered support and wielded growing influence. In 1976, West Germany's new Nature Protection Law was passed, and already at the end of the 1970s, local urban nature protection laws and initiatives were introduced, influenced by the new scientific studies on urban ecosystems. Here West Berlin was again a pioneer with its Nature Protection Law of 1979.

Urban ecological research increased in the 1970s, applying proven ecological methods to urban conditions. Cities such as Berlin were now recognised as highly complex ecosystems. The basic theoretical assumption was that in the city the various utilisation forms and activities had a fundamental impact on plants and animals and their communities:

> Within the settled areas there are primarily utilisation forms which are dominating the pattern and distribution of organisms. The basis of nature protection in the city is, therefore, to analyse systematically the most important types of land-use and to describe species content and ecological characteristics. In the final result it becomes clear, which land use forms are particularly poor in species and require management for the re-implementation of nature.[7]

This, and other statements, illustrated that land-use was now seen as the key factor in applied urban ecological research and urban nature conservation. A new research direction and its application, urban habitat mapping, focused on detailed land-use mapping and the inventory of ecological factors, that is on the compilation of 'land-use types' or 'urban structural types'.[8] In this way, structural categories in urban landscape were used as the principal guidelines and categories for urban ecological and geographical studies in the 1970s and 1980s.[9] Mapping guidelines and categories were developed for a broad, comparative application in urban and non-urban nature protection.[10] However, the adoption of land-use types as ecological units

---

7   H. Sukopp, W. Kunick and Ch. Schneider, 'Biotopkartierung im besiedelten Bereich von Berlin (West): Teil II: Zur Methodik von Geländearbeit', *Garten und Landschaft* 7 (1980), 565; K.-H. Hülbusch, 'Landschaftsökologie der Stadt: Naturschutz und Landschaftspflege zwischen Erhalten und Gestalten', Referate und Ergebnisse des Deutschen Naturschutztages 1982 vom 19.–23. Mai 1982 in Kasse, *Jahrbuch. f. Naturschutz u. Landschaftspflege*, Bonn 33 1982, 38–61.
8   F. Duhme and Th. Lecke, 'Zur Interpretation der Nutzungstypenkarte München', *Landschaft und Stadt* 18, (1986), 174–185.
9   F. Duhme and S. Pauleit, 'Naturschutzprogramm für München: Landschaftsökologisches Rahmenkonzept', *Geogr. Rundsch.* 44/10 (1992), 554–561; K. Reidl, 'Flora und Vegetation als Grundlage für den Naturschutz in der Stadt. Teil 1: Methodik und Ergebnisse der Kartierung am Beispiel Essen, *Naturschutz und Landschaftsplanung* 4 (1992), 36–141.
10 Arbeitsgruppe Methodik der Biotopkartierung im besiedelten Bereich (ed.), 'Flächendeckende Biotopkartierung im besiedelten Bereich als Grundlage einer ökologisch bzw.

Figure 7.3 Protected landscapes (wetlands, flood plains forests and other forests) in the Landscape Development Plan of the city of Halle/Saale. Source: Stadt Halle (ed.), Landschaftsentwicklungsplan. Halle/Saale 1995.

was implemented without detailed investigations into the functions and processes of the human impacts on specific land-use forms and spaces.

Following on from this, new management practices of urban nature, including urban green space, were increasingly seen as necessary for urban nature protection. This required a methodological framework for nature protection in the city. The methodology of a *Kartierung schutzwürdiger Biotope in Bayern*[11] [The Mapping of Biotopes with Conservation Value in Bavaria] was applied to cities. This urban methodology was pioneered by the ecologist Herbert Sukopp and his school in Berlin, who developed the idea of managing urban biotopes in the framework of a complete overview

am Naturschutz orientierten Planung: Grundprogramm für die Bestandsaufnahme und Gliederung des besiedelten Bereichs und dessen Randzonen', *Natur und Landschaft* 61/10 (1986), 371–389; Arbeitsgruppe Methodik der Biotopkartierung im besiedelten Bereich (ed.), 'Flächendeckende Biotopkartierung im besiedelten Bereich als Grundlage einer am Naturschutz orientierten Planung: Programm für die Bestandsaufnahme, Gliederung und Bewertung des besiedelten Bereichs und dessen Randzonen', *Natur und Landschaft* 68/10 (1993), 491–526.

11 G. Kaule, 'Kartierung schutzwürdiger Biotope in Bayern. Erfahrungen 1974', *Verh. Gesell. f. Ökol.* 3 (1975), 257–260.

of all the natural elements in a city. Biotope mapping (as a tool for both research and management) was implemented in a similar form in nearly all other German federal states of the Federal Republic of Germany between 1975 and 1985, and soon spread to other countries. The central goals were the protection of species, and scenic (landscape) values, as well as of landscape functions. This classified the protection often according to the very distinctive elements of urban landscapes, such as wetlands and floodplain forests.[12]

For nature protection in the city, the land-use patterns were used as an initial reference point. The land-use types associated with human activities were also useful for understanding nature protection in urban conditions. Nature protection in cities and towns had to change its paradigm from one of protecting threatened species of plants and animals to a more general one of protecting nature and the urban landscape. The task of urban nature protection now embraced protecting organisms and habitats because of their meaning for the relationship of urban dwellers and their natural environment.[13]

The method in the late 1970s of mapping 'habitats which are of conservation value' applied the principle of selective habitat mapping to cities.[14] Increasingly this was no longer restricted to an overview of all land-use structures. Surveys based on floral and vegetation science used structurally homogeneous areas and a complete monitoring of the urban landscape.[15]

Land-use based habitat mapping became the principal method for habitat-mapping of urban landscapes based on a standardised method in 1986.[16] This programme was again revised in 1993.[17] Traditional urban nature protection methodologies were extended by addressing a number of new topics, such as: possibilities for acceptance of different types of urban nature;

---

12 Breuste, 'Nutzung als Untersuchungsgegenstand'; J. Breuste and S. Wohlleber, 'Goals and measures of nature conservation and landscape protection in urban cultural landscapes of Central Europe – examples from Leipzig', in Breuste, Feldmann and Uhlmann (eds), *Urban Ecology,* 676–682.

13 Sukopp, Kunick and Schneider, 'Biotopkartierung im besiedelten Bereich von Berlin (West)', 565–569; H. Sukopp and P. Weiler, 'Biotopkartierung im besiedelten Bereich der Bundesrepublik Deutschland', *Landschaft und Stadt* 18/1 1986, 25–38; H. Sukopp und L. Trepl, Naturschutz in Großstädten. Unpublished manuscript, Berlin: 1990.

14 M. Brunner, F. Duhme, F. Mück, J. Patsch and F. Weinisch, 'Kartierung erhaltenswerter Lebensräume in der Stadt', *Gartenamt* 28 (1979), 1–8; N. Müller and R. Waldert, 'Erfassung erhaltenswerter Lebensräume für Pflanzen und Tiere in der Stadt: Augsburg Stadtbiotopkartierung', *Natur und Landschaft* 56/2 (1981), 419–429.

15 H. Sukopp, W. Kunick and Ch. Schneider, 'Biotopkartierung in der Stadt', *Natur und Landschaft* 54/3 (1979), 66–68; W. Kunick, Stadtbiotopkartierung Berlin-Kreuzberg Nord (Unpublished manuscript, Techn. Univ. Berlin 1978).

16 Arbeitsgruppe Methodik der Biotopkartierung im besiedelten Bereich (ed.), 'Flächendeckende Biotopkartierung im besiedelten Bereich als Grundlage einer ökologisch bzw. am Naturschutz orientierten Planung: Grundprogramm für die Bestandsaufnahme und Gliederung des besiedelten Bereichs und dessen Randzonen', *Natur und Landschaft* 61/10 (1986), 371–389.

17 Arbeitsgruppe Methodik der Biotopkartierung im besiedelten Bereich (ed.), 'Flächendeckende Biotopkartierung im besiedelten Bereich als Grundlage einer am Naturschutz orientierten Planung: Programm für die Bestandsaufnahme, Gliederung und Bewertung des besiedelten Bereichs und dessen Randzonen', *Natur und Landschaft* 68/10 (1993), 491–526; Breuste, 'Nutzung als Untersuchungsgegenstand'.

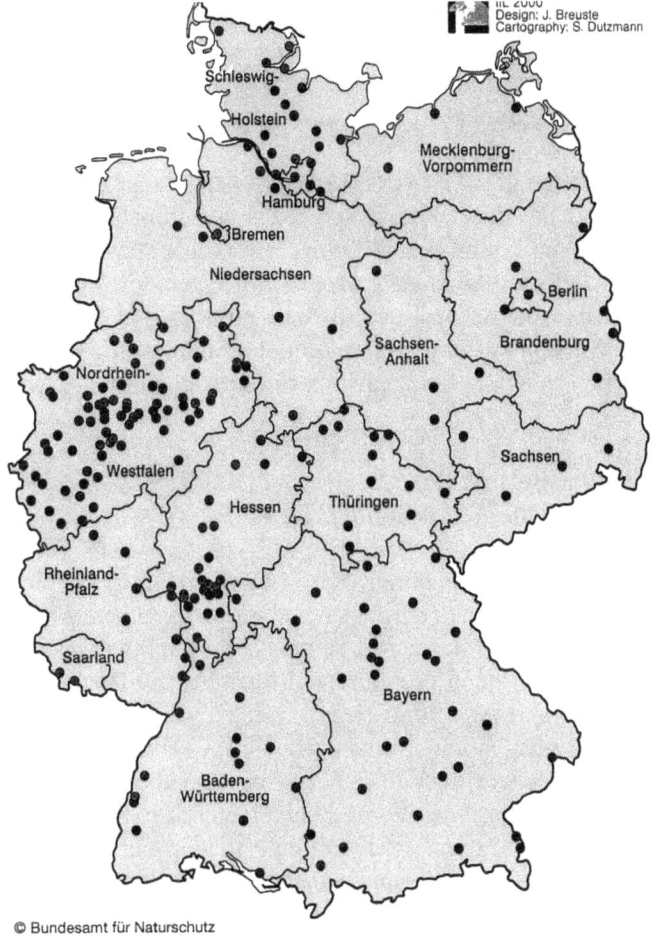

Figure 7.4 Habitat mapping in Germany. Source: J. Breuste, 'Nutzung als Untersuchungsgegenstand und Raumbezug der Stadtökologie', *Natur und Landschaft* 33 (2/3) (2000), 95–100.

non-restricted utilisation of space by children; recreation and environmental education; the production of fruit, vegetables and ornamental plants in urban surroundings; environmental protection (water regime, climate, air hygiene, noise pollution); and monitoring of environmental changes and loads. All of this involves ecological research about the complex role of green spaces.[18]

Despite these newly recognised objectives, nature protective habitat-mapping was often applied as rather narrow species and biotope protection up to the 1980s. By contrast, more recently urban 'nature' has been considered the prime concern and an educational resource. This has opened a new agenda in Germany: according to this, urban green spaces have to be

---

18 H. Sukopp and P. Weiler, 'Biotopkartierung im besiedelten Bereich der Bundesrepublik Deutschland', *Landschaft und Stadt* 18/1 1986, 25–38.

improved to better serve social and cultural functions, and to work in the teaching and nature education of young people. The provision of ecosystem services has become a new theme of investigation, and also a priority of planning and management aspects of German cities during the last few years.[19]

The role of urban green spaces for environmental education expanded in the 1990s. After the first Urban Ecological Educational Trail was built in 1987, in Brunswick, more than 40 other such trails were established. All these trails included urban green spaces of conserved or restored nature and were furnished with explanations (panels, pictures and so on).[20] Special attention was given to the needs of schools. Some schools already have green spaces for practical field teaching (for example in Leipzig).

Multi-functional green spaces: developments of the last two decades

Analysis of the multiple functions of green spaces which also incorporate socio-spatial, ecological, planning and demographic parameters is recognised in many present-day German cities. Several cities are involved in European Union funded urban green programmes (for instance, Bochum, Dresden, Leipzig and Berlin).[21] An interdisciplinary catalogue of criteria (ICC) was developed for the evaluation of urban green space services and functions in the European Union project URGE[22], and implemented in German cities. The ICC is seen as a toolbox of criteria and methods to make planning and management of urban green spaces more efficient and appropriate for a multi-disciplinary development and management of urban green spaces. The 'toolbox' contains a set of criteria and methods to support the categorisation of green spaces, problems, planning and decision-making processes on the scale of the whole city or urban region. The ICC established a selection of criteria to describe the present status of urban green; it also provides guidelines and methods to evaluate urban green functions; to address planning targets; consider the impact of disturbances or new assets; and examine strategic paths for the development and planning of actions. Thus the ICC

---

19 A. Schulz, 'Der KÖH-Wert, Modell einer komplexen, planungsrelevanten Zustandserfassung', *Inform. z. Raumentwicklung* (1982), 847–863; R. Wittig and R. Schreiber, 'A quick method for assessing the importance of open spaces in towns for urban nature conservation', *Biological Conservation* 26/1 1983, 57–64; Trepl, 'Flora und Vegetation'; W.M. Kaerkes, 'Zur ökologischen Bedeutung urbaner Freiflächen – dargestellt an Beispielen aus dem mittleren Ruhrgebiet', in *Materialien zur Raumordnung*, 35, Bochum 1987; B. Fischerlehner, 'Naturerleben bei Kindern', in Bayerisches Landesamt für Umweltschutz München, Inst. f. Botanik u. Pharmazeutische Biologie der Univ. Erlangen (eds). *13. Jahrestagung "Biotopkartierung im besiedelten Bereich"*. Erlangen 1992, 95–102.
20 W. Schulte and C. Hettwer, 'Lehrpfade und Naturerlebnispfade zur Dorf- und Stadtökologie in Deutschland', *Natur und Landschaft* 74/1, 3–10.
21 A. Werquin, C., B. Duhem, G. Lindholm, B. Oppermann, S. Pauleit and S. Tjallingii (eds), *Green Structure and Urban Planning*. European Commission COST Action C11, Final report, 2005.
22 URGE-Project (ed.), *Urge-Project*, 2007,

	ICC – Interdisciplinary Catalogue of Criteria		Site Level	
URGE urban green environment	Checklist criteria at Site Level			

1. Quantity of urban green space			Sucessfully applied?	
Criterion		Indicator	YES	NO
1.1	Surface area of urban green space	Surface area of green space	☐	☐
1.2	Extent of edge effects	Shape index	☐	☐
1.3	Isolation from other green spaces	Inter-patch distance to nearest neighbouring patch	☐	☐
1.4	Connectivity to other green spaces	Presence of different types of green corridors that link a site to other urban green spaces	☐	☐
1.5	Soil sealing	Proportion of soil surface with disturbed drainage/water dynamics	☐	☐
1.6	Integration in green system	Existence and effectiveness of instruments capable of integrating the green space to a green system	☐	☐

2. Quality of urban green space			Sucessfully applied?	
Criterion		Indicator	YES	NO
2.1	Biodiversity	Species diversity: number of rare/threatened/endangered species	☐	☐
		Biotope diversity: the number of different habitat types found within the green site	☐	☐
2.2	Surface disturbance	Proportion of the surface which is heavily worn	☐	☐
2.3	Naturalness	Proportion of indigenous/exotic, rare (threatened/endangered) and protected species in relation to the total number of species found at the site	☐	☐
		The proportion of native biotopes to the total surface area of the site	☐	☐
2.4	Pollution	Soil pollution and quality	☐	☐
		Air quality	☐	☐
		Water quality	☐	☐
		Noise attenuation	☐	☐
		Urban intrusion	☐	☐
2.5	Regulatory effects	Leaf Area Index expressed as an average for the city	☐	☐
		Volume of vegetation	☐	☐
		Proportion of the total area, which is in shade on a sunny summer day at noon	☐	☐
		Wind amelioration	☐	☐
2.6	Aesthetic value	Statements from local residents with regard the aesthetic value of the green space	☐	☐
		Added price and rental value of houses/offices	☐	☐

Figure 7.5 Interdisciplinary Catalogue of Criteria to evaluate urban green space functionality (ICC). Source: URGE-Project, 2007.

presents a holistic approach to the analysis of urban green spaces and the development of guidelines for their management. This criteria catalogue has already been successfully approved in several German and other European cities and is a step forward to giving urban green space a necessarily central position in German urban planning.[23]

23 URGE-Project (ed.), *Urge-Project*, 2007

Conclusion

After the Second World War, green space development followed variable trends in East and West Germany. The first decades were characterised by two tendencies: the establishment of green spaces in newly built urban areas, during the reconstruction of German cities; and the overbuilding of urban green by other forms of land use in densely built-up areas. Both tendencies developed in parallel. Ambitious planning concepts such as green networks and green corridors could not be realised, and remained incomplete in most cities. This was mainly due to insufficient political attention to the issue in the context of urban planning, and inadequate financing. During the early decades after the war, the concept of urban greening was mainly focused on the recreational function of green spaces. This included the dedication of green spaces to sports activities, especially in East Germany.

The 1970s brought an ecological revolution in German society and a fast growing awareness of urban environmental problems. Green spaces became the subject of ecological research and of ecologically oriented management, such as biotope mapping. The ecological significance of urban green spaces dominated the scientific literature and influenced urban green space management from the 1970s to the end of the 1990s.

In step with this development, a new perspective on urban green spaces as a multifunctional type of land use began to evolve in the late 1980s and has dominated subsequent research and decision-making. Green spaces are important components of urban renewal and function as planning tools for the improvement of residential estates in the 1990s in former East German cities. Here experiments have been undertaken with new forms of green spaces (not only the old-style parks), such as urban wilderness areas, short-term use of urban wasteland, and allotment gardens, albeit as before within the constraints of inadequate public financing. No less important, a more social interpretation of urban green spaces, as opposed to a purely ecological or aesthetic discourse, has come to shape urban green space planning and management in Germany at the start of the twenty-first century.

JARMO SAARIKIVI[1]

8 Ecological Perspectives of Sports Green Space in the Contemporary European City

Nowadays, when the majority of the world's human population lives in cities, one could argue that accelerated urbanisation is just another piece of evidence that the human economy is 'decoupling' itself from the environment, that mankind is finally leaving nature and the rural countryside behind. The massive migration of people to cities produces a dramatic shift in the spatial distributions of human populations, but this shift occurs without reducing the material dependence of people on biophysical products and ecosystem services. Even as we urbanise, human beings remain dependent on the environment, and the cities become the major drivers of global ecological change.[2]

Many people see urbanisation as a transition to a higher level of civilisation. Western industrial society tends to associate 'rural' and 'agricultural' with general underdevelopment and presumptively inferior peasant culture. But the close (emotional) connection with nature is what is left of our origins in hunting and gathering. Socio-cultural evolution has not destroyed the importance of natural environments for our well-being.[3] Instead these values act, for example, as determinants in the choice of one's residential location.[4] Some migration studies in central Europe already report a continuous outflow from cities to peri-urban regions. One of the main reasons for the urban sprawl in Belgium is the lack of public green spaces and children's play grounds in the cities.[5]

This holds true also in Helsinki, where some higher income families with children are attracted from the city to small-scale housing in greener living

---

1   I am grateful for funding for this work to the European Studies Network project *Green Space, Sport and the City*.
2   W. Rees, 'The conundrum of urban sustainability', in D. Devuyst, L. Hens and W. De Lannoy (eds), *How Green is the City? Sustainability Assessment and the Management of Urban Environments*. New York: Columbia University Press 2001, 37–41.
3   Rees, 'The conundrum of urban sustainability'.
4   L. Tyrväinen, *Monetary Valuation of Urban Forest Amenities in Finland*. Finnish Forest Research Institute, Research papers 739, 1999.
5   E. Pelfrene, *In-en uitwijking in Vlaamse steden en gemeenten. Analyse naar leeftijd en ruimtelijke structuren voor de periode 1996–1998*, Stativaria nr. 24, juni 2000. Brussel: Ministerie van de Vlaamse Gemeenschap 2000.

environments in nearby communities.[6] At the same time in many cities in Europe, including Helsinki, a major decrease in urban green space is a future fact.[7] The declining quality and quantity of urban green space has consequences for nature and biodiversity in cities. Urban green spaces are becoming more homogenous and they risk losing their habitat and species heterogeneity.[8] This is detrimental to both biodiversity itself and to residents enjoying nature in cities. It is a political matter, whether the urban green spaces have a high priority in improving the quality of urban life or whether they are just seen as a cheap medicine for those less fortunate inhabitants, unable to choose their residential location.

The aim of this paper is to review published works on the ecology of urban green space with an emphasis on sports grounds, especially golf courses. In particular, I discuss the question of how the different animal groups are affected by the increasing number of golf courses and the growing popularity of golf in recent decades. Also of interest are other golf related environmental issues. This knowledge can aid golf course management and conservation efforts in identifying the environmentally important key elements behind the processes of creating and maintaining a golf course.

Diversity

Although people have played their part in enhancing biological diversity in the past, at the present time we are evidently contributing to a reduction in the diversity of the natural environment.[9] It is therefore no wonder, that there has been an increasing concern about the magnitude of global biodiversity loss,[10] especially in urban areas, that were previously often seen as anti-life and therefore not worth studying with regard to ecology. It was assumed that few plants or animals could survive in an urban setting and that urban animal and plant communities were products of coincidence.[11] That attitude began to change some 30 years ago, thanks to the increasing number of ecological studies of cities.[12] These studies shared the astonished realisation that environments created by humans provide habitats for characteristic species and that these species reoccur under similar conditions. Analyses have shown that urban areas include a wide variety of habitats,

---

6 *Väestönmuutokset Helsingissä 1997–2000*. Helsinki: Helsingin kaupungin tietokeskus 2002.
7 S. Pekkala, 'Migration in a core-periphery model: analysis of agglomeration in regional growth centres', paper presented at the 40th European congress of the Regional Science Association, Aug. 29-Sep. 1. 2000, Barcelona, Spain. *ERSA conference papers*, 2000.
8 O.L. Gilbert, *The Ecology of Urban Habitats*. London: Chapman and Hall 1989.
9 E-L. Hallanaro and M. Pylvänäinen, *Nature in Northern Europe – Biodiversity in a Changing Environment*. Copenhagen: Nordic Council of Ministers 2002.
10 K.J. Gaston, *Biodiversity: A Biology of Numbers and Differences*. London: Blackwell Science 1996.
11 H. Sukopp, 'Urban ecology – scientific and practical aspects', in J. Breuste, H. Feldmann and O. Uhlmann, (eds), *Urban Ecology*. Berlin: Springer 1998, 3–16.
12 J. Niemelä, 'Is there a need for a theory of urban ecology?', *Urban Ecosystems* 3 (1999), 57–65; J. Niemelä, 'Ecology and urban planning', *Biodiversity and Conservation* 8 (1999), 119–131.

organisms and communities.[13] When looking closer at the European cities and their wide range of open spaces – parks of different types, public and private gardens, schoolyards, playgrounds, churchyards, allotment gardens, sports grounds and so forth – it is clear that these areas are occupied by a wide variety of flora and fauna, often more diverse in urban contexts than in other habitats within the same region.[14] For example, a close examination of the 50 granite steps of Helsinki's Lutheran Cathedral revealed the presence of 74 plant species.[15] And a study in Riga showed that the greatest levels of species diversity for epiphytic mosses and lichens have been recorded in the city's 18 cemeteries, where as many as ten different moss species may be found on a single gravestone.[16] Furthermore, rare and endangered species sometimes occur in urbanised habitats and thus could be conserved there.

On the other hand, urbanisation is a threat to many natural habitats and species.[17] For instance, in the German city of Munich over 180 plant species have become locally extinct in the past 100 years.[18] The possibly deceptive richness of species in cities results from the diverse mosaic nature of urban habitat composition, anthropogenic patterns and processes, and unique urban features like high levels of air pollution, disturbance intensity, the heat island phenomenon and the presence or greater abundance of exotic species.[19] Although most of the factors which affect urban ecosystems also operate in non-urban areas, the combination and extreme intensity of these factors means that unique ecosystems develop with species combinations peculiar to urban areas.[20] Therefore we may conclude that urbanisation affects biodiversity by providing resources that may be beneficial to some species, typically generalist or pioneer species, and by limiting the resources for other species that typically favour stable and disturbance-free ecosystems. Urbanisation also creates new unique ecosystems and thereby supports many species characteristic of the city. Furthermore, urban biodiversity plays an important educational role in providing opportunities for observing wildlife

---

13 Sukopp, 'Urban ecology'.
14 P. Clark and M. Hietala, 'Helsinki and green space 1850–2000: an introduction' in P. Clark (ed.), *The European City and Green Space: London, Stockholm, Helsinki and St Petersburg*. Aldershot: Ashgate 2006, 175–187.
15 A. Kurtto and L. Helynranta, 'Helsingin kasveja 2: Erään "kansallisnäkymän" kasvisto' (English abstract: Plants of Helsinki 2: Flora of a 'national scene'), *Lutukka* 13 (1997), 56.
16 L. Liepina, Institute of Biology, University of Latvia, in E-L. Hallanaro and M. Pylvänäinen, *Nature in Northern Europe – Biodiversity in a Changing Environment*. Copenhagen: Nordic Council of Ministers 2002, 281.
17 T. Kendle and S. Forbes. *Urban Nature Conservation*. London: Chapman and Hall 1997; S. Godefroid, 'Temporal analysis of the Brussels flora as an indicator for changing environmental quality', *Landscape and urban planning* 52 (2001), 203–224; Niemelä, 'Is there a need for a theory of urban ecology?'.
18 F. Duhme and S. Pauleit, 'A landscape ecological masterplan for the city of Munich', in J.O. Riley and S.E. Page (eds), *Habitat Creation and Wildlife Conservation in Urban and Post-Industrial Environments*. Chichester: Packard Publishing 2000.
19 M.J. McDonnell, S.T.A. Pickett, P. Groffman, P. Bohlen, R.V. Pouyat, W.C. Zipperer, R.W. Parmelee, M.M. Carreiro and K. Medley, 'Ecosystem processes along an urban-to-rural gradient', *Urban Ecosystems* 1 (1997), 21–36.
20 Sukopp, 'Urban ecology'.

and enjoying nature. It sensitises humans to green spaces and natural systems which balance the 'inorganic' reality of buildings and urban infrastructure.

Green space

The presence of nature within the city can take many forms. Natural systems tend to persist in the urban setting where strong geological features (that is, rivers and geological contours) and land not suitable for building exist. Basic concepts instrumental in guiding the development of the city are also important factors, given that the density of buildings and infrastructure leaves either more or less room for nature. In certain cities, zoning regulations call for plenty of open space, which is usually developed as green space.

The conventional ornamental park was previously seen as a relic of an age when man felt it necessary to proclaim his ability to subdue the wilderness.[21] It was a costly anachronism, unsuitable for 'active' recreation and too intensively managed to be of much value to wildlife. Instead of these stereotypes, we are nowadays offered the prospect of a richer landscape in which advantage is taken of existing features of scenic and biological merit, and an attempt is made to achieve ecological diversity by making use of indigenous trees, shrubs and herbs. Some cities provide semi-natural open spaces which retain much of their biological interest, in spite of urban growth and intensive public use. Varied attitudes to open space have shaped the city's recreational areas with leisure activities, such as sport, having a big impact on the creation and use of green space.[22]

Sport and green space

Sports areas comprise a significant part of the European urban landscape and are in some cities more extensive than nature reserves.[23] Potentially, they are a major ecological resource. The growth of sports, particularly organised competitive sports, has had a large impact on the urban environment since the First World War. There has been a major expansion of sports grounds both as multi-functional parks and as specialist sites.[24] As many parks and sports areas have been created in areas of open space that once supported semi-natural habitat, this habitat often persists in fragments within the parks. If sufficiently large, these relic sites can provide the resource from which more extensive areas of grassland or woodland can be restored or created. However, in many countries, there is a lack of ecological knowledge

---

21 I.C. Laurie, *Nature in Cities: The Natural Environment in the Design and Development of Urban Green Space*. London: Wiley 1979.
22 Clark and Hietala, 'Helsinki and green space 1850–2000'.
23 Clark and Hietala, 'Helsinki and green space 1850–2000'.
24 P. Clark, J. Saarikivi and S. Jokela, 'Nature, sport and the European city', (paper at a conference on 'Nature in the City', German Historical Institute, Washington, December 2005), to be published in S. Dümpelmann and D. Brantz, (eds), *The Place of Nature in the City* (forthcoming).

that should be considered in urban planning.[25] The biodiversity of urban habitats is still poorly documented in many cities and the potential of urban green space in preserving nature and wildlife is often not recognised or appreciated.

Playing fields in particular are often regarded as inimical to wildlife, because of the need for very regular mowing. However, even these fairly featureless 'green deserts' can support, for example, a variety of common bird species.[26] Formal parks (and amenity open space) tend to support a wider range of biodiversity, because they have a greater degree of structural diversity (for instance, trees and shrubberies are scattered throughout the mown grassland) and many support a diversity of habitats including ponds, lakes and copses. The most significant threat to the biodiversity of formal parks, sport pitches and amenity open space is unenlightened management.[27] The extensive area of these types of open space provides enormous potential for habitat enhancement and habitat creation within the limits imposed by the needs of formal recreational and amenity areas. Where there are no remnants of former habitats, habitat creation techniques can be applied to make new habitats (such as ponds or wildflower meadows). Alternatively the existing park maintenance regime can be amended to allow greater structural diversity. Relaxing mowing regimes, cutting hedges less frequently or delaying the removal of accumulated leaf litter are some options.

Recreation and leisure activities are not without their environmental impacts, however. Various kinds of disturbance to the wildlife and habitats have been reported.[28] The effect of transport, noise and pollution are all concerns expressed by governmental bodies.[29] On the other hand, a significant opportunity for awareness-raising about biodiversity is offered by the popularity of parks and open spaces. Far more people are likely to visit their local park or playing field than their local nature reserve. Providing information about the biodiversity of a local park is the first step in promoting a greater appreciation of biodiversity in general.

---

25 Niemelä, 'Ecology and urban planning'.
26 M. Merola-Zwartjes and J.P. DeLong, 'Southwestern golf courses provide needed riparian habitat for birds', *USGA Turfgrass and Environmental Research Online* 4 (2005), 1–18.
27 S. Carruthers, J. Smart, T. Langton and J. Bellamy. *Open Space in London: Habitat Handbook 2*. London: The Greater London Council 1986; D. Dawson and A. Worrell, *The Amount of Each Kind of Ground Cover in Greater London*. London: London Ecology Unit 1992.
28 S.A. Boyle and F.B. Samson, 'Effects of non-consumptive recreation on wildlife: a review', *Wildlife Society Bulletin* 13 (1985), 110–116; N. Chettri, E. Sharma and D.C. Deb, 'Bird community structure along a trekking corridor of Sikkim Himalaya: a conservation perspective', *Biology Conservation* 102 (2001), 1–16.
29 R.P. Cincotta, J. Wisnewski and R. Engelman, 'Human population in the biodiversity hotspots', *Nature* 404 (2000), 990–992; J. Mikola, M. Miettinen, E. Lehikoinen and K. Lehtilä, 'The effects of disturbance caused by boating on survival and behaviour of velvet scooter *Melanitta fusca* ducklings', *Biology Conservation* 67 (1994), 119–124; D. Sun and D. Walsh, 'Review of studies on environmental impacts of recreation and tourism in Australia', *Journal of Environmental Management* 53 (1998), 323–338; R.A. Tanner and A.C. Gange, 'Effects of golf courses on local biodiversity', *Landscape and Urban Planning* 71 (2004), 137–146.

Golf courses from an ecological point of view

Worldwide, there are over 30,000 golf courses and 55 million people who play golf. Many countries are currently experiencing a golf boom.[30] New courses are being built at a rate that makes golf course development one of the fastest growing types of land development in the world. For example, in the United States, on average, more than one new golf course opens every day (509 new courses opened in 1999).[31] This is not happening without environmental impacts, and golf's effect on the environment is a hotly debated topic in many countries.[32] In fact, golf is a sport that has been placed at the environmental spotlight, because it occupies such large areas of land – often close to urban centres where human welfare, ecosystem health, water quality and other issues are of paramount importance.[33] A typical 18 hole golf courses comprises 54 hectares of land and some 70 percent of that area is considered rough, or out-of-play areas, that has potential for creating significant wildlife benefits.[34]

The public perception of golf courses is overwhelmingly that they are bad for the environment, although this depends on whether one is actively involved with the game. In a survey of 400 people in south-east England during 2002, it was found that 80 per cent of respondents who played golf answered that courses were good for the environment, while among non-players, this figure fell to 36 per cent. Among players, the most common reason given for courses being beneficial was that they preserve areas of natural habitat. However, among non-players, the most common reason for courses being seen as detrimental was that they destroyed areas of natural habitat! This clear disparity of views shows that many people may be misinformed about the value of golf courses from an environmental point of view. The survey also showed that there is much anti-golf feeling amongst the general public.[35]

Historically, golf courses have been seen as places of heavy pesticide and fertilizer use, but there are few scientific studies done on their impact.[36] Habitat modification, chemical contamination, water management and

---

30 M.R. Farrally, A.J. Cochran, D.J. Crews, M.J. Hurdzan, R.J. Price, J.T. Snow and P.R. Thomas, 'Golf science research at the beginning of the twenty-first century', *Journal of Sports Sciences* 21 (2003), 753–765; M.R. Terman, 'Natural links: naturalistic golf courses as wildlife habitat', *Landscape and Urban Planning* 38 (1997), 183–197.
31 S. Thien, S. Starrett, R. Robel, P. Shea, D. Gourlay and C. Roth, 'A multiple index environmental quality evaluation and management system: application to a golf course', *USGA Turfgrass and Environmental Research Online* 3 (2004), 1–10.
32 R.G. Dodson, *Managing Wildlife Habitat on Golf Courses*. New York: John Wiley & Sons 2000.
33 P. Stangel and K. Distler, 'Golf courses for wildlife: looking beyond the turf', *USGA Turfgrass and Environmental Research Online* 1 (2002), 1–8.
34 D. Tilly, 'Golf course program produces many birds', *The Journal of the North American Bluebird Society,* Winter ed. (2000); Terman, 'Natural links'.
35 A.C. Gange, D.E. Lindsay and M.J. Schofield, 'The ecology of golf courses', *Biologist* 50 (2003), 63–68.
36 M. Yasuda and F. Koike, 'Do golf courses provide a refuge for flora and fauna in Japanese urban landscapes?', *Landscape and Urban Planning* 75 (2006), 58–68.

urbanisation around golf courses are all concerns that have been expressed by those who claim that golf courses are a poor option ecologically.[37] However, until recently, there was little hard evidence to confirm whether golf courses were good or bad for the environment at the landscape scale. What information there is suggests that golf courses are not significant sources of water pollution and may be equal to many natural habitats in terms of animal and plant diversity.[38] Previously, there were only a handful of research studies that employed a strict scientific method to the study of wildlife on golf courses.[39] Many recent studies have shown that golf courses compare well in terms of wildlife abundance and diversity to that of adjacent areas of land. A feature of these studies is that the diversity of taxa on golf courses has been compared with areas of pristine natural habitat.[40] As shown by Gange and Lindsay (2002), a more realistic question to ask, in terms of landscape ecology, is how the biological diversity of a golf course compares with that of the habitat from which the course was constructed, as land targeted for golf development in the last 20 years has been almost exclusively farmland.[41] Tanner and Gange (2004) conclude that golf courses of any age can enhance the local biodiversity of an area by providing a greater variety of habitats than intensively managed agricultural areas.[42]

Another important aspect is that there is much that golf courses can do to enhance the quality of habitats they possess and that ecologists can help course managers to maximise the conservation potential of the specific habitats. Nowadays, many golf courses actively promote nature conservation and harbour some of the rarest plant and animal species. Furthermore, golf courses can now participate in environmental management programmes and their efforts are being recognised through a national awards programme in the UK.[43] In recent years, there has also been an attempt to increase the provision of various open and semi-open recreational areas in cities. In

---

37 Terman, 'Natural links'; E.A. Murphy and M. Aucott, 'An assessment of the amounts of arsenical pesticides used historically in a geographical area', *Science of the Total Environment,* 218 (1998), 89–101; S. Cohen, A. Svrjcek, T. Durborow and N.L. Barnes, 'Groundwater and surface-water risk assessments for proposed golf courses', *ACS Symposium series* 522 (1993), 214–227; M.C. Markwick, 'Golf tourism development, stakeholders, differing discourses and alternative agendas: the case of Malta', *Tourism Management* 21 (2000), 515–524; A.E. Platt, *Toxic Green: The Trouble with Golf.* Washington DC: Worldwatch Institute, 1994, 1–6.
38 S. Cohen, A. Svrjcek, T. Durborow and N.L. Barnes, 'Water quality impacts on golf courses', *Journal of Environmental Quality* 28 (1999), 798–809; A.C. Gange and D.E. Lindsay, 'Can golf courses enhance local biodiversity?', in E. Thain (ed.), *Science and Golf IV.* (Proceedings of the World Scientific Congress on Golf) London: Routledge 2002, 721–736; Terman, 'Natural links'.
39 R.B. Blair, 'Land use and avian species diversity along an urban gradient', *Ecological Applications* 6 (1996), 506–519; B.H. Green and I.C. Marshall, 'An assessment of the role of golf courses, in Kent, England, in protecting wildlife', *Landscape and Urban Planning* 14 (1987), 143–154; Terman, 'Natural links'; M.R. Terman, 'Ecology and golf: saving wildlife habitats on human landscapes', *Golf Course Manage* 68 (2000), 183–197.
40 Blair, 'Land use and avian species diversity'; Terman, 'Natural links'.
41 Gange and Lindsay, 'Can golf courses enhance local biodiversity?'.
42 Tanner and Gange, 'Effects of golf courses on local biodiversity'.
43 Gange, Lindsay and Schofield, 'The ecology of golf courses'.

Scotland the total area of land managed for formal recreation, such as playing fields and golf courses, more than doubled between 1951 and 1991.[44] Nowadays a significant trend in golf course management is to create more naturalistic landscapes and there are organisations whose aim is the promotion of environmental sustainability in golf. Evenso, conservationists fear that the 'environmentally friendly' label will be used to justify building golf courses in ecologically sensitive areas.[45]

Diversity of habitats in golf courses

One vital factor limiting biodiversity is the availability of suitable habitats. Whether it is a square meter of turf, a suitable habitat for many insects or spiders, or a forest patch or a pond, all of these habitats can often be found on a golf course. In fact, golf course landscapes typically contain diverse habitats, ranging from ponds to streams, wetlands to grasslands, and woodlands to mature forests. This variety of habitats provides unique opportunities for wildlife.[46]

Availability of water is often the most important factor in attracting wildlife. Even the smallest pond will soon be occupied by variety of insects and other invertebrates. And a riparian habitat may act as an oasis for a great diversity of birds and other wildlife. Generally, increased habitat heterogeneity will also increase the total species richness of an area. From that perspective, golf courses have great potential in preserving biodiversity, since they usually contain a diverse range of habitats and microhabitats and varying degrees of land use intensity and human influence.

In many Northern countries golf courses are used for golf for only about half of the year or even less. During the winter, the courses are covered with snow and very little activity takes place in those areas. Then again in summertime, golf courses may represent dry, warm and sunny open areas – a habitat otherwise rare in northern latitudes. These habitats are especially important environments for many rare insect and vascular plant species.[47]

The habitat patches in golf courses are usually relatively small. Fragment area and isolation are often the strongest landscape descriptors of wildlife distribution and abundance. Many species are sensitive to fragmentation and the number of species usually reduces as the patches become smaller and more isolated.[48] In golf courses, however, isolation is less intense than

---

44 E.C. Mackey, M.C. Sherwy and G.J. Tudor, *Land Cover Change: Scotland from the 1940s to the 1980s*. Edinburgh: The Stationery Office 1998.
45 M.J. Santiago and A.D. Rodewald, 'Considering wildlife in golf course management', Ohio state university extension fact sheet 2004; Anon. 'Golf course design has become more environmentally friendly', *U.S. Water News Online*.
46 Santiago and Rodewald, *Considering Wildlife in Golf Course Management*.
47 S. From, 'The significance of dry and sunny habitats for biodiversity', in S. From (ed.) *The Ecology and Threatened Species of Sunny and Dry Habitats*. Finnish Environment Institute 2005, 7–11.
48 See for example, J.S. Gray, 'Effects of environmental stress on species rich assemblages', *Biological Journal of the Linnean Society* 37 (1989), 19–32.

in areas isolated by, for example, concreted asphalt or buildings. In addition, the body size of animals is often correlated with habitat patch size: smaller animals manage to survive in smaller habitat patches, whereas larger ones require larger habitats.[49] In theory, the ecological role of golf courses, and many other smaller habitat parcels, may be to serve as stepping stones for animals like dispersing birds or as corridors between different natural areas or as buffer zones between nature and the urban infrastructure.[50] Thus, even the smallest habitat patch can be valuable in preserving our present capital of biodiversity.

Birds in golf courses

Of the wildlife at golf courses, birds are probably the most studied group of animals. They are relatively easy to monitor and the ecology, behavior and habitat needs for many species are well known. Many recent studies on wildlife in golf courses have focused on birds, and some papers and a book have been published recently.[51] These studies show that a golf course can provide suitable habitats for many bird species and, in some cases, can even contain birds that are of conservation concern. Recently golf courses have played an important role in conservation efforts for species like the eastern bluebird (*Sialia sialis*), tree swallow (*Tachycineta bicolor*), purple martin (*Progne subis*), red-cockaded woodpecker (*Picoides borealis*) and osprey (*Pandion haliaetus*).[52] This is not to suggest that golf courses can fill the ecological roles of natural landscapes or support the biodiversity present in indigenous ecosystems, but they may provide specific habitat components for some declining habitat specialists.[53]

In Northern America, populations of the previously common red-headed woodpecker (*Melanerpes erythrocephalus*) have declined because of the loss of its natural habitat (oak savanna, farmland and other open habitats with trees), but the species has found its way to the highly modified habitats

---

49 See for example, D.A. Saunders, R.J. Hobbs and C.R. Margules, 'Biological consequences of ecosystem fragmentation: a review', *Conservation Biology* 5 (1991), 18–32.
50 Terman, 'Natural links'.
51 D.A. Cristol and A.D. Rodewald, 'Introduction: can golf courses play a role in bird conservation?', *Wildlife Society Bulletin* 33 (2005), 407–410; S. Dale, 'Effects of a golf course on population dynamics of the endangered ortolan bunting', *Journal of Wildlife Management* 68 (2004), 719–724; D.H. Gordon, S.G. Jones and G.M. Phillips, 'Golf courses and bird communities in the South Atlantic coastal plain', *USGA Turfgrass and Environmental Research Online* 2 (2003), 1–9; Merola-Zwartjes and DeLong, 'Southwestern golf courses'; P.G. Rodewald, M.J. Santiago and A.D. Rodewald, 'Habitat use of breeding red-headed woodpeckers on golf courses in Ohio', *Wildlife Society Bulletin* 33 (2005), 448–453; M.D. Smith and C.J. Conway, 'Use of artificial burrows on golf courses for burrowing owl conservation', *USGA Turfgrass and Environmental research online* 4 (2005), 1–6; C.L. White and M.B. Main, 'Habitat value of golf course wetlands to waterbirds', *USGA Turfgrass and Environmental Research Online* 3 (2004), 1–10; S.W. Gillihan, *Bird Conservation on Golf Courses: A Design and Management Manual*. Chelsea: Ann Arbor Press 2000.
52 Tilly, 'Golf course program'.
53 Cristol and Rodewald, 'Introduction'.

of golf courses, where old, large trees suitable for nesting, together with open woodland, still occur.[54] Another declining open habitat species, the burrowing owl (*Athene cunicularia*) has also benefited from the open foraging areas, lower predation rate and even artificial burrows in golf courses in south-central Washington.[55]

However, there are also species that try to avoid golf courses and species that are negatively affected by them. Long term monitoring of the endangered ortolan bunting (*Emberiza hortulana*) in Norway clearly showed that the species do not find golf courses attractive. Although calling males were present at a golf course, the low pairing success and other indicators suggests that the females avoided the golf course, indicating the habitat's unsuitability for breeding.[56]

The diverse habitat types provided by many golf courses have the potential to offer valuable habitats for many bird species. Many birds gather in golf courses, because of the irrigation systems that provide water. Also wetlands attract birds and other wildlife to golf courses. As natural wetlands decline in availability and quality, alternative habitats such as created wetlands may become increasingly important to wetland dependent wildlife. Many water birds have benefited from the artificially created water places in golf courses as foraging, nesting or resting (wading species) grounds. The extent to which water birds use golf course ponds is primarily related to pond size, ability of the birds to access prey, and habitat features that influence security and foraging success. Therefore some habitat modifications could benefit water birds, and thus provide management options for the golf course. Providing a diversity of habitat features would offer the greatest benefits to the largest number of species and thus increase the value of golf course ponds to water birds.[57]

Golf courses can support a greater number of birds than surrounding natural areas, a response that is common throughout studies of avian responses to urbanisation.[58] The overall impact of a golf course on avian community composition appears to be very different in different types of environments. Loss of bird diversity is likely when development occurs in an area that had an initially high diversity of habitats. It should be noted, however, that even when the overall species richness and diversity increase in a golf course, the original and native bird community may suffer.[59] Furthermore, a large

54 Rodewald, Santiago and Rodewald, 'Habitat use of breeding red-headed woodpeckers'.
55 Smith and Conway, 'Use of artificial burrows on golf courses'.
56 Dale, 'Effects of a golf course on population dynamics'.
57 White and Main, 'Habitat value of golf course wetlands to waterbirds'
58 Merola-Zwartjes and DeLong, 'Southwestern golf courses'; S.R. Beissinger and D.R. Osborne, 'Effects of urbanization on avian community organization', *Condor* 84 (1982), 75–83; J.T. Emlen, 'An urban bird community in Tuscon, Arizona: Derivation, structure, regulation', *Condor* 76 (1974), 184–197; R.J. Green, 'Native and exotic birds in a suburban habitat', *Australian Wildlife Resources* 11 (1984), 181–190; E. Hohtola, 'Differential changes in bird community structure with urbanization: a study in central Finland', *Ornis Scandinavica* 9 (1978), 94–99.
59 Blair, 'Land use and avian species diversity'; R.B. Blair and A.E. Launer, 'Butterfly diversity and human land use: species assemblages along an urban gradient', *Biological Conservation* 80 (1997); Terman, 'Natural links'.

number of birds that are relatively widespread and abundant may even be considered pests or nuisance species (for example starlings, *Sturnus vulgaris*).[60] In Finland, many species of birds have been observed at golf courses, because courses are often situated on the migratory routes of the birds. Species are easy to notice and recognise in open areas and golfers also enjoy watching birds on their outings and may prefer to frequent golf courses that attract large numbers of birds. Recently the dramatic increase in Barnacle Goose (*Branta leucopsis*) populations has evoked entirely different range of emotions. These large herbivorous birds are attracted by urban areas that offer the resources and refuge geese need. Birds take advantage of the lush lawns at golf courses, while their droppings and wing molt feathers annoy players. Methods like shooting and harassment by dogs are sometimes used to control them.

Several factors such as food, the availability of suitable nest sites, and inter-specific competition have been recognised to be important in determining avian habitat selection and community structure.[61] Lack of vegetation, especially standing or felled dead trees, reduces the availability of foraging or nesting areas even in moderately urbanised habitats, whereas winter feeding and nest boxes may help species that are able utilise these resources. Many cavity-nesting species suffer from the lack of suitable nest sites in urban areas and by providing nest boxes for them, their numbers and reproduction success may be increased. For example, in our studies in the Helsinki region, over 40 per cent of the 200 nest boxes for passerine birds placed in the urban golf courses were soon occupied and used for breeding.

Amphibians in golf courses

Increasing concern about amphibian populations has captured global attention, because of well-documented declines at local, regional and even global scales.[62] A variety of factors have been suggested to be behind these declines, for instance introduced predators, fertilizers, pollutants and UV-B radiation.[63] One of the leading factors for the decrease of pond-breeding

60 Merola-Zwartjes and DeLong, 'Southwestern golf courses'.
61 J. Jokimäki and E. Huhta, 'Artificial nest predation and abundance of birds along an urban gradient', *Condor* 102 (2000), 838–847; P. Osborne and L. Osborne, 'The contribution of nest site characteristics to breeding-success among blackbirds *Turdus merula*', *Ibis* 122 (1980), 512–517.
62 See for example, R.A. Alfrod and S.J. Richards, 'Global amphibian declines: a problem in applied ecology', *Annual Review of Ecology and Systematics* 30 (1999), 133–165; A.R. Blaustein and D.B. Wake, 'Declining amphibian populations: a global phenomenon?', *Trends in Ecology and Evolution* 5 (1990), 203–204; J.E. Houlahan, C.S. Findlay, B.R. Schmidt, A.H. Meyer and S.L. Kuzmin, 'Quantitative evidence for global amphibian population declines', *Nature* 404 (2000), 752–755.
63 P.S. Corn, 'Amphibian declines: review of some current hypothesis', in D.W. Sparling, C.A. Bishop and G. Linder (eds), *Ecotoxicology of amphibians and reptiles*. Society of Environmental Toxicology and Chemistry. Pensacola: Setac Press 2000, 663–696; A.R. Blaustein and J. M. Kieseker, 'Complexity in conservation: lessons from the global decline of amphibian populations', *Ecology letters* 5 (2002), 597–608; M. Pahkala, K. Räsänen, A. Laurila, U. Johanson, L.O. Björn and J. Merilä, 'Lethal and sublethal effects of UV-B/pH synergism on common frog embryos', *Conservation Biology* 16 (2002), 1063–1073.

amphibians is the impact of habitat fragmentation.[64] Amphibians have complex life cycles that make them vulnerable to habitat loss and fragmentation. They also have permeable skin that makes them exceptionally sensitive to changes in microclimate and microhabitat composition.

Typically, adults of most of the species migrate annually to their breeding ponds, where they spend less than one month and then migrate again to their non-breeding territory. Ponds are often used only for mating and deposition of eggs and by larvae during their development to metamorphosis. Adults are usually highly site faithful to their breeding pond, returning to the same pond year after year, whereas metamorphoses tend to disperse across the landscape and often breed in new ponds.[65] Typically, amphibians decrease in richness with increasing urbanisation due to loss of habitat, declining water quality and the increase of predator populations,[66] but there are some studies that indicate that amphibians may show high levels of species diversity in urban areas.[67]

A few empirical studies have been published about the impacts of golf course design and management on amphibian populations.[68] Golf course wetlands can support amphibian species that have lost valuable wetland habitat elsewhere in urban areas and, if developed and maintained wisely, they may have the potential to function as mini-preserves for even some of the rarest and most susceptible species. Generally, amphibians seem to prefer ponds that dry annually (to avoid fish), are not very sensitive to grass height around ponds, are more likely to disperse through forested landscapes and often avoid moving across broad expanses of turf, such as fairways or greens.[69] Furthermore, amphibians are sensitive to even low concentrations of many pesticides (insecticides, fungicides, herbicides and so on) during the breeding season, most likely because some compounds accumulate in the jelly layers of amphibian eggs.[70] Although these chemicals are not always directly lethal, the decreased hatching rates or slower growth rates can be

---

64 R.M. Lehtinen, S.M. Galatowitsch and J.R. Tester, 'Consequences of habitat loss and fragmentation for wetland amphibian assemblages', *Wetlands* 19 (1999), 1–12.
65 P.W.C. Paton and R.S. Egan, 'Strategies to maintain pond-breeding amphibians on golf courses', *USGA Turfgrass and Environmental research online* 1 (2002), 1–7.
66 H.-J. Mader, 'Animal habitat isolation by roads and agricultural fields', *Biological Conservation* 29 (1984), 81–96; M.J. Rubbo and J.M. Kiesecker, 'Amphibian breeding distribution in an urbanized landscape', *Conservation Biology* 19 (2005), 504–511; C.C. Vos and J.P. Chardon, 'Effects of habitat fragmentation and road density on the distribution pattern of the moor frog *Rana arvalis*', *Journal of Applied Ecology* 35 (1998), 44–56.
67 M.J. Castro, J.M. Oliveira and A. Tari, 'Conflicts between urban growth and species protection: can midwife toads (*Alytes obstetricans*) resist the pressure?' in N. Ananjeva and O. Tsinenko (eds) *Herpetologia Petropolitana*. Proceedings of the 12th ordinary general meeting of the Societas Europaea Herpetologica, August 12–16, 2003, St. Petersburg, Russian Journal of Herpetology 12 (Suppl. 2005), 126–129.
68 See for example, J.H. Howard, E.J. Shannon and J. Ferrigan, 'Golf course design and maintenance: impacts on amphibians', *USGA Turfgrass and Environmental Reseach Online* 1 (2002), 1–21; Paton and Egan, 'Strategies to maintain pond-breeding amphibians'.
69 Paton and Egan, 'Strategies to maintain pond-breeding amphibians'.
70 For example, D.W. Sparling, G.M. Fellers and L.L. McConnell, 'Pesticides and amphibian population declines in California, USA', *Environmental Toxicology and Chemistry* 20 (2000), 1591–1595.

damaging to the population over the years.[71] Decreased fitness in amphibians in golf courses has been demonstrated even in colour pattern asymmetry as a correlate of habitat disturbance.[72] Casual observation that certain amphibian species thrive in strongly manipulated environments can lead to erroneous conclusions regarding long term impacts. In order to maintain populations of amphibians in golf courses, attention must be paid to ecosystem integrity, where ponds, movement corridors and habitats during both the breeding and non-breeding seasons, must be considered. Seasonal wetlands appear to be very effective in enhancing amphibian diversity on golf courses.[73]

Lack of suitable breeding areas affects amphibians as well as birds or other wildlife in urban areas. Although amphibians exhibit breeding philopatry or site fidelity, they are likely to colonise new ponds or other breeding areas relatively quickly. In Helsinki, the new man-made water bodies in golf courses are likely to be colonised by the common frog (*Rana temporaria*) and smooth newt (*Triturus vulgaris*) within a couple of years after construction.

Mammals in golf courses

The scene of golfers enjoying a game on a beautiful day can be remarkably improved by having a herd of deer standing alert in the distant background. This image makes clear an association many link to golf: a unique opportunity to spend time outside, in close contact with nature, while pursuing a challenging sport.[74] But how real is that connection between golf and the natural world, and can that connection be made stronger to the benefit of both the golf community and nature?

Golf courses make excellent habitats for many mammal species. However, the damage these animals cause to the carefully manicured turf grass is unsightly and may be expensive to repair. Therefore many mammal species are not welcomed to golf courses and great efforts are taken to control them.

In her work with small mammal diversity on golf courses, Barthelmess (2004) showed that species richness was similar on the golf course and in woodland patches. However, there were differences in abundances of certain species. Diurnal squirrels were abundant in golf courses, whereas nocturnal squirrels and jumping mice were more common in the woodland sites. Certain species appeared intolerant of the golf course landscape. The golf course was not the equivalent of natural woodland in supporting small mammal

---

71 Howard, Shannon and Ferrigan, 'Golf course design and maintenance'.
72 A.N. Wright and K.R. Zamudio, 'Color pattern asymmetry as a correlate of habitat disturbance in spotted salamanders (*Ambystoma maculatum*)', *Journal of Herpetology* 36 (2002), 129–133.
73 B.S. Metts, D.E. Scott and W. Gibbons, Enhancing amphibian biodiversity on golf courses through the use of seasonal wetlands. Poster. Savannah River Ecology Laboratory, the University of Georgia; Aiken 2002.
74 E.L. Barthelmess, 'Managing golf courses for small mammal diversity', *USGA Turfgrass and Environmental Reseach Online* 3 (2004), 1–10.

communities.[75] Managers can do a lot to encourage mammal species on golf courses. Hanging bat boxes can provide needed nesting and shelter areas for these nocturnal mammals. Bats in turn can be an effective natural tool for pest control and thus reduce the volume of pesticides required.[76]

Invertebrates in golf courses

The impact of invertebrates on the wildlife in golf courses is similar to the natural habitats. Many invertebrates have important functional roles as selective herbivores and pollinators and as prey for a variety of bird species. However, some species like earthworms or ants can be regarded as nuisance or pests when they occur in large numbers and produce castings or mounds on particularly low-cut turf like putting greens, approaches and collars, tee boxes and fairways. Great efforts are taken to control or manage these animals on golf courses and since they are widely considered beneficial organisms due to certain favourable attributes, such as soil formation, aeration and drainage, organic matter breakdown and incorporation, and even enhancement of microbial activity, their control is often unnecessary in other areas than low-cut turf. It is in fact illegal to use any pesticide applications to control earthworms in some countries (in the USA, for example) and therefore alternative, non chemical management strategies are needed and used.[77] Furthermore soil living organisms can be difficult to control. For example, conventional chemical control often fails to control the turf-infesting ants, because insecticides usually can not eliminate the subterranean queen.[78]

Animals that do not disrupt the smoothness and uniformity of the golf course or do not otherwise impact on the game, are usually well tolerated by both players and golf course managers. Therefore invertebrates that are relatively small and seldom seen receive little attention and public interest. Butterflies, however, have received some attention, since some private golf courses (in Florida) have participated in the conservation of the endangered Schaus Swallowtail (*Papilio aristodemus ponceanus*). The project, which involved reintroductions and improving and restoring habitats, has been successful in conserving the species and bringing invertebrate conservation to the forefront.[79] In their work to assess the conservation value of golf courses for butterflies Porter et al. (2004) recorded on-site vegetation and used GIS –methodology and aerial photography to estimate percentages of

---

75 Barthelmess, 'Managing golf courses'.
76 Santiago and Rodewald, *Considering Wildlife in Golf Course Management*.
77 R.M. Maier and D.A. Potter, 'Nuisance ants on golf courses', *USGA Turfgrass and Environmental research online* 2 (2003), 1–5; R.C. Williamson and S. Hong, 'Managing earthworm castings in golf course turf', *USGA Turfgrass and Environmental Research Online* 3 (2004), 1–6.
78 R. Lopéz, D.W. Held and D.A. Potter, 'Management of a mound-building ant, *Lasius neoniger*, on golf putting greens and tees using delayed–action baits or Fipronil', *Crop Science* 40 (2000), 511–517.
79 J.C. Daniels and T.C. Emmel, 'Florida golf courses help save the endangered butterfly', *USGA Turfgrass and Environmental Reseach Online* 3 (2004), 1–7.

different land-cover types surrounding the golf courses. They found out that butterfly abundance was significantly related to land-cover characteristics and that the abundance, species richness and diversity measures decreased as the surrounding land uses of the course urbanised. A golf course with larger areas of natural land cover, along with a larger buffer comprised of patches of natural habitats, improves its conservation value. Thus the landscape surrounding courses has a significant influence on their ability to support diverse butterfly communities.[80]

In our work to investigate the insect diversity and community structure in urban golf courses in the Helsinki region, we used Carabid beetles (Coleoptera: Carabidae) as bio-indicators. Carabid beetles have previously been used as appropriate indicators in different types of environmental and ecological studies in which the focus has been on biodiversity,[81] disturbance gradients,[82] trampling,[83] grassland[84] – or forest management,[85] landscape changes,[86] habitat classification[87] and conservation. A common trend is that large, poorly dispersing specialist species decrease with increased disturbance, while small generalist species with good dispersal ability increase.[88] Our data of 72 species and 6941 individuals indicate that the carabid communities are variable between the golf courses although almost identical microhabitats can be found in every course. This suggests that the age, maintenance, human intensity or landscape level parameters play an important role in structuring carabid beetle and possibly also other insect communities.

---

80 E.E. Porter, D.N. Pennington, J. Bulluck and R.B. Blair, 'Assessing the conservation value of golf courses for butterflies', *USGA Turfgrass and Environmental research online* 3 (2004), 1–13.
81 J. Niemelä, J. Kotze, A. Ashworth, P. Brandmayr, D. Konjev, T. New, L. Penev, M. Samways and J. Spence, 'The search for common anthropogenic impacts on biodiversity: a global network', *Journal of Insect Conservation* 4 (2000), 3–9.
82 J. Kotze and J. Niemelä, 'A global network for examining anthropogenic impacts on carabids: first results'. in Szyszko et al. (eds), *How to Protect What We Know about Carabid Beetles.* Warsaw Agricultural University Press 2000, 155–169; S. Venn, J. Kotze and J. Niemelä, 'Urbanization effects on carabid diversity in boreal forests', *European Journal of Entomology* 100 (2003), 73–80.
83 A.-C. Grandchamp, J. Niemelä and J. Kotze, 'The effects of trampling on assemblages of ground beetles (Coleoptera, Carabidae) in urban forests in Helsinki, Finland', *Urban Ecosystems* 4 (2000), 321–332; S. Lehvävirta, J. Kotze, J. Niemelä, M. Mäntysaari and B. O'Hara, 'Effects of fragmentation and trampling on carabid beetle assemblages in urban woodlands in Helsinki, Finland', *Urban Ecosystems* 9 (2006), 13–26.
84 A.-C. Grandchamp, A. Bergamini, S. Stofer, J. Niemelä, P. Duelli and C. Scheidegger, 'The influence of grassland management on ground beetles (Carabidae, Coleoptera) in Swiss montane meadows', *Agriculture, Ecosystems and Environment* 110 (2005), 307–317.
85 M. Koivula and J. Niemelä, 'Boreal carabid beetles (Coleoptera, Carabidae) in managed spruce forests – a summary of Finnish case studies', *Silva Fennica* 36 (2002), 423–436.
86 J. Niemelä, 'Carabid beetles (Coleoptera: Carabidae) and habitat fragmentation: a review', *European Journal of entomology* 98 (2001), 127–132; K. F. Davies and C.R. Margules, 'Effects of habitat fragmentation on carabid beetles: experimental evidence', *Journal of Animal Ecology* 67 (1998), 460–471.
87 M.D. Eyre and M.L. Luff, 'A preliminary classification of European grassland habitats using carabid beetles', in: N. Stork (ed.), *Ground beetles: Their role in Ecological and Environmental studies*. Andover: Intercept Publications, 1990, 227–236.
88 J. Rainio and J. Niemelä, 'Ground beetles (Coloptera: Carabidae) as bioindicators', *Biodiversity and Conservation* 12 (2003), 487–506.

Vegetation in golf courses

Much of the reduction in species richness is obviously caused by the loss of vegetation. The number of species of animal taxa, such as birds and insects, tends to correlate with the number of plant species in an area.[89] Of importance also is the origin of the vegetation. Historically, native plants have often been ignored in favour of exotic species in landscaping, but this philosophy is gradually changing. Public gardens using native species are being built, books are being written on the subject and a number of landscape architects are emphasising the use of native plants for many types of landscapes.[90] There is no doubt that native vegetation will play an important role in our environmental policy of the twenty-first century, although the critical discussion about 'the native plant enthusiasm' is called for and the approach has been criticised from the historical perspective with arguments referring to Nazi politics.[91] Nevertheless, the policy of native plants can be applied also in golf course management. If necessary, fairly large areas of golf courses can easily be left in a natural state with native plants growing not only in the roughs and out-of-play areas, but also in natural buffer zones that partially surround the courses, separating them from nearby residential housing.

Plants on golf courses absorb carbon dioxide, release oxygen and filter pollutants from runoff. It is beneficial to choose combinations of plants that will offer year-round colour: berries, pollen, seeds and nuts. Clustered vegetation will be aesthetically pleasing, and wildlife will benefit from the resulting corridors and buffers which link the habitat patches scattered throughout the golf course. Favouring native species is generally good practice. Native plants are less costly to maintain since they are well suited to regional moisture and soil conditions, therefore generally requiring less fertilizers and irrigation. In addition, choosing plants that are naturally resistant to pests and disease will make sustaining healthy specimens easier, without relying on chemical pesticides and herbicides.[92]

Landscape vegetation can be selected also according to on its food value. Food resources are very important in attracting wildlife and an appropriate choice of vegetation can provide food resources for a variety of wildlife species. Similarly important is vegetation that provides cover and protection for animals to carry out activities functions such as breeding, nesting, resting and travel.[93]

---

89 M. McKinney, 'Urbanization, biodiversity and conservation', *BioScience* 52 (2002), 883–890.
90 G.L. Hightshoe, *Native trees for Urban and Rural America*. Iowa State University Press 1978; G. Gröning and J. Wolschke–Bulmahn, 'Some notes on the mania for native plants in Germany', *Landscape Journal* 11 (1992), 125; G. Gröning and J. Wolschke–Bulmahn, 'The native plant enthusiasm: ecological panacea or xenophobia?', *Landscape Research* 28 (2003), 75–88.
91 J. Weston, 'Using native plants in the golf course landscape', *USGA Green Section Records*. 28 (1990), 12–16; Gröning and Wolschke–Bulmahn, 'Some notes on the mania for native plants'; F. Uekötter, 'Native plants: a Nazi obsession?', *Landscape Research* 32 (2007), 379–383.
92 Santiago and Rodewald, *Considering Wildlife in Golf Course Management*.
93 P. Bronski, 'Walk on the wild side: creating a wildlife sanctuary on your golf course', *California Fairways*. sept/oct 2002.

Conclusion

Our environment is under pressure like never before. Development and urban sprawl, coupled with population explosion, is placing ever-increasing demands on natural resources such as water, timber, energy and land. Wildlife is left with fewer and fewer places to turn to as these demands encroach on their native habitats. Among the many human activities that cause habitat loss, urban development produces some of the greatest local extinction rates and frequently eliminates the large majority of native species.[94] Also, urbanisation is often more lasting than other types of habitat loss.[95] The great conservation challenge of urban growth is that it replaces native species with abundant non native species. This replacement contributes to the process of biotic homogenisation of cities that threatens to reduce the biological uniqueness of local ecosystems.[96]

Undisturbed, pristine habitats are commonly the choice of ecologists in search of knowledge about the machinery of nature. However, with the spread of urbanisation, it is time for ecologists to study human-dominated landscapes such as golf courses or urban green space. Transforming a golf course into a wildlife sanctuary is not just about wildlife, though. It is an opportunity to improve the quality of the course, and foster positive community relations. Most important of all, it means that we will be protecting environmental quality and ensuring that our precious and valuable natural resources will continue to sustain us well into the future. From forests to deserts, mountains to valleys, wetlands to meadows, the landscape and our golf courses, are a mosaic that supports an incredible diversity of wildlife. This diversity of ecosystems and wildlife presents an exciting opportunity for golf courses. Every course has the chance to make a significant contribution to the environment given the broad range of habitats that can be found throughout the different regions.[97] The general ethical consideration that arises from that perspective is whether the small habitat patches within the human dominated landscape are worthwhile?

---

94 B. Czech, P.R. Krausman and P.K. Devers, 'Economic associations among causes of species endangerment in the United States', *BioScience* 50 (2000), 593–601; T.R. Vale and G.R. Vale, 'Suburban bird populations in west–central California', *Journal of Biogeography* 3 (1976), 157–165; M. Luniak, 'The development of bird communities in new housing estates in Warsaw', *Memorabilia Zoologica* 49 (1994), 257–267; I. Kowarik, 'On the role of alien species in urban flora and vegetation', in P. Pysek, K. Prach, M. Reymánek and P.M. Wade (eds), *Plant invasions – General Aspects and Special Problems*. Amsterdam: SPB Academic 1995, 85–103; J.M. Marzluff, 'Worldwide urbanization and its effects on birds', in J.M. Marzluff, R. Bowman and R. Donnely (eds), *Avian Ecology in an Urbanizing World*. Norwell: Kluwer 2001, 19–47; McKinney, 'Urbanization, biodiversity and conservation'.
95 McKinney, 'Urbanization, biodiversity and conservation'.
96 J. Olden, L. Poff, M. Douglas and K. Fausch, 'Ecological and evolutionary consequences of biotic homogenization'. *Trends in Ecology & Evolution* 19 (2003), 18–24; J. Lockwood and M. McKinney (eds), *Biotic Homogenization*. New York: Kluwer Academic/Plenum Publishers 2001; R.B. Blair, 'Birds and butterflies along urban gradients in two ecoregions of the U.S', in J.L. Lockwood and M.L. McKinney (eds), *Biotic Homogenization*. New York: Kluwer 2001, 33–56.
97 Bronski, 'Walk on the wild side'.
98 McKinney, 'Urbanization, biodiversity and conservation'.

The answer is yes. Urban planners should always find ways to preserve biodiversity as cities expand outward and subsequently modify natural habitat. Such efforts would most likely focus on preserving as much remnant natural habitat as possible. And where intensive land development has already occurred, native biodiversity can be increased by a variety of currently popular techniques. But none of the conservation efforts can be applied without a proper knowledge of ecological principles and urban ecosystems. This is where the study of urban ecology can serve and show its potential. And if the impacts of urbanisation on native ecosystems are to be reduced, one should also consider the human population as a part of the urban ecosystem. A second way in which the study of urban ecology can serve conservation is by helping to develop a more ecologically informed public. In fact, this could be the most important application of urban ecology.[98] A more ecologically informed public could greatly improve the social support for conservation of native species in all ecosystems.

However, our knowledge for managing a diversity of small habitats in a region is incomplete.[99] We are relatively well informed about the various ways in which urban (and suburban) expansion harms native ecosystems, but we know very little on how to promote (native) species conservation in urban ecosystems or about restoring modified habitats. We know various ways to measure diversity and are quite aware of series of common pitfalls that underlie in quantifying and comparing taxon richness.[100] Richness or biodiversity are values that need further interpretation in order to be valuable in context of human life. Actions needed to manage cities as living spaces for people and nature involve all aspects of life that can alter human behaviour. In that context, scientific research is one important component as an applied method of influencing and directing human behaviour.

---

99 Terman, 'Natural links'.
100 McKinney, 'Urbanization, biodiversity and conservation'; A. Magurran, *Ecological Diversity and its Measurement*. London: Croom Helm 1988; N. Gotelli and R. Colwell, 'Quantifying biodiversity: procedures and pitfalls in the measurement and comparison of species richness', *Ecology letters* 4 (2001), 379–391.

SUVI TALJA[1]

9 Golf and Green Space in Finland: Tali Golf Course Since 1932

This chapter explores the social and environmental impacts of golf courses in the twentieth-century city. So far in Finland relatively little work has been done on this type of green space either by historians or social scientists. The significance of different urban green space, whether golf courses or parks and recreation grounds, needs to be examined from both a social and an environmental perspective. From the social point of view, urban green spaces can be evaluated according to their accessibility and to the opportunities they offer for recreation and aesthetic enjoyment. From an environmental point of view, green spaces are deemed important to sustain biodiversity in urban areas. Importantly, these aspects are not unrelated: a rich biodiversity enhances the recreational value of parks and other green space, while heavy usage may in turn cause a deterioration of the environment.[2]

Concerns about the social and environmental impacts of golf courses started to surface worldwide in the 1980s. From a social sustainability standpoint, golf courses came to be regarded as problematic since they occupy large land areas for private club usage in contested urban areas.[3] A golf course is a good example of a cultural landscape, which is constructed through

---

1 I am grateful for funding for this work to the European Studies Network project *Green Space, Sport and the City*.
2 R. Pakarinen, 'Luonto ja virkistys', in L. Autio (ed.), *Katsaus Helsingin ympäristön tilaan 2003*. Helsingin kaupungin ympäristökeskuksen monisteita 9/2003, 32. See also discussion on urban green space and its ecosystem services and accessibility issues in E. Andersson, 'Urban landscapes and sustainable cities', *Ecology and Society* 11(1):34 (2006). [online] URL: http://www.ecologyandsociety.org/vol11/iss1/art34/. For integrating information from ecological and social systems for urban planning purposes, particularly in the Finnish context, see V. Yli-Pelkonen and J. Niemelä, 'Linking ecological and social systems in cities: Urban planning in Finland as a case', *Biodiversity and Conservation* 14 (2005), 1947–1967.
3 T. Suoninen, R. Porttikivi, A. Särkioja, and I. Taipalinen, *Tarinaharjun golfkentän pinta- ja pohjavesivaikutukset: Loppuraportti*, Suomen ympäristö 590. Kuopio: Pohjois-Savon ympäristökeskus 2002, 7; Golfkenttäselvitys. Vantaan kaupunki. Yleiskaavatoimikunta 24.4.2002, oheisaineisto 6/1. Kaupunkisuunnitteluyksikkö C9: 2003 – KSY 5/2003. Conflicting situations may emerge between a golf course and the surrounding society, especially in areas with scarce land resources. See for example M.C. Markwick, 'Golf tourism development, stakeholders, differing discourses and alternative agendas: the case of Malta', *Tourism Management* 21 (2000), 515–524.

negotiations between different interest groups, exercising power over one another.[4] In Finland, the large green spaces in urban areas are usually forests, or other 'relict' nature left between built-up areas. However, there is also 'man-made' nature such as built parks, owned and maintained by the public sector, which are generally open for public use. In addition, the public right of access with some restrictions, 'the everyman's right', to private land is in effect.[5] In this context, golf courses occupying large contiguous land areas are quite a unique type of green space in Finnish cities. In Helsinki, golf courses are the only large green spaces that mainly serve private use.[6]

In terms of the environmental problems of golf courses, a rather heated discussion has developed during recent decades. Arguably, the environmental problems associated with golf course development include the loss of natural habitats, while regular golf course maintenance, such as irrigation, may divert water resources as well as leach fertilizers and pesticides.[7] Environmental aspects attracted official recognition in Europe in the mid-1990s, when the *European Golf Association (EGA)* founded its Ecology Unit, and produced environmental guidelines for golf clubs throughout Europe. The *Golf Union of Finland* established a programme for environmental impacts evaluation in 1995, and two years later it went on to give advice to golf clubs on environmental matters and establish guidelines for them. By the end of the 1990s, 40 out of 91 golf courses in Finland had created an environmental plan in accordance with the *Union's* guiding principles.[8]

This chapter examines social and environmental issues relating to golf courses through a case study of the Tali course, the oldest and best known in Finland. In the first part, I examine the spatial growth of the Tali course since its establishment in 1932 and the discussions the area has provoked since then. One of the concerns is to see when questions regarding the social impacts of the golf course gained importance. In the second part, I discuss in detail the users of the golf course and its interaction with adjacent areas. The last part focuses on environmental aspects, looking specifically at the evolution of golf course maintenance. For a more detailed discussion of the relationship between ecology and golf courses see Chapter Eight by Jarmo

---

4   J. Bale, *Landscapes of Modern Sport*. Leicester: Leicester University Press 1994, 11. Bale has used the concept of 'territoriality' from R.D. Sack (*Human territoriality: Its Theory and History*. Cambridge: Cambridge University Press 1986.) to discuss power relations in the sports context.
5   Yli-Pelkonen and Niemelä, 'Linking ecological and social systems', 1948; E. Pouta and M. Heikkilä (eds), *Virkistysalueiden suunnittelu ja hoito*. Helsinki: Ympäristöministeriö, alueidenkäytön osasto 1998, 3.
6   Arguably, restrictions to public use are also found in other recreational and sports spaces in Helsinki, such as fenced football grounds with well-maintained grass lawns, fenced allotment garden areas with restricted visiting hours for public, and marines for private boating clubs.
7   Suoninen, Porttikivi, Särkioja and Taipalinen, *Tarinaharjun golfkentän pinta- ja pohjavesivaikutukset*, 10–14.
8   T. Tynjälä, 'Nurmon Ruuhikoskelle tuorein ympäristöohjelma', *Etelä-Suomen Sanomat* 20.11.1998; Suoninen, Porttikivi, Särkioja and Taipalinen, *Tarinaharjun golfkentän pinta- ja pohjavesivaikutukset*, 8; Golfliitto, Golfkentän ympäristöjärjestelmä, <http://www.golf.fi/portal/golfliitto/kentanhoito/ymparistonsuojelu/golfkentan_ymparistokasikirja/golfkenttien_ymparistotyo>, Visited 9.1.2009.

Saarikivi. Before examining the Tali golf course in detail, it is necessary to set it in context, by looking at the history of golf in Finland.

Golf culture in Finland

The origins of golf culture in Finland go back to the 1930s, when the Tali golf course was established by an association called *Helsingin Golfiklubi ry – Helsingfors Golfklubb rf*. For a long time, only a few people played golf in Finland, and golf was perceived as an elite sport. The *Finnish Golf Union* was founded in 1957 to serve as a national umbrella-organisation for the handful of existing local clubs: *Helsingin Golfiklubi ry* (established in 1932), *Viipurin Golf ry* (1938), *Porin Golfkerho ry* (1939) and *Kokkolan Golf ry* (1957). At that time, the total number of the Finnish golf club members was only about 600. Golf culture started relatively late in Finland compared to the other Nordic countries. In Norway, golf had been played from the 1920s, and there were three courses in 1949. In Sweden, the game had gained popularity even earlier. The first golf club was founded in Gothenburg at the turn of the twentieth century, and by the end of the 1940s the country had already 4,000–5,000 golfers.[9] Part of the explanation for the tardy development in Finland may be the less developed economy and lower income levels than in the other Nordic countries.

It was not until the 1980s that golf started gaining in popularity among the wider public in Finland. The number of full-size golf courses grew along with an increase of smaller golf courses. While the Tali golf course was still the only full-size course with 18 holes at the beginning of the 1980s, by 1987 the number had grown to nine. The golf boom was partly stimulated by a corresponding economic boom in Finland. Club membership experienced the fastest growth in the middle of the 1980s. At the same time, the pattern of course ownership changed. Until the early 1980s golf courses had been managed by voluntary associations, but in the 1980s property developers and private companies got interested in the golf business. However, compared to England and Germany (see Chapter Ten by James and Gardner and Chapter Eleven by Eisenberg and Tamme) commercial companies have become only minor actors in Finnish golf culture. By the mid-1990s most new golf courses were commercial, share-based companies, but the shares were mainly owned by golfers, and the companies were not concerned to make profits. An association was usually established to run practical matters at the course. In 2004, only 15 out of the 110 golf courses in Finland were owned by profit-making companies. The most common type of golf course was one combining an association and a company owned by golfers.[10]

---

9 L. Tilander, *Vuodesta 1932 Helsingin Golfklubi*. Helsinki: Helsingin Golfklubi 2007, 24; 'Golfiharrastus kasvamassa pohjoismaissa', *Uusi Suomi* 10.9.1949.
10 R. Vahtokari, 'Golfin suunta on citius, altius, fortius', *Suomen kuvalehti*,38/1985; K. Kyheröinen. 'Ilon ja murheen greenit', *Suomen kuvalehti 26–27/1987;* 'Miljonääri takaa uuden golfkentän', *Talouselämä,* 32/2006, 35–40; S. Tiainen, Golftoiminnan kansantaloudelliset vaikutukset, Opinnäytetyö, Helsingin liiketalouden ammattikorkeakoulu 2006, 7–8.

Finnish municipalities have not been very active in the provision of golf courses, and by 2004 only 0.4 per cent of the building costs of the Finnish courses had been covered by public funds. The first municipal golf course was founded at Jyväskylä in 1978, but its management was subsequently given over to an association. Today, there are no municipal golf courses in Finland. In Sweden, one can see a very different system. Swedish municipalities established golf courses already in the 1930s, and today, 85 per cent of the country's golf courses are 'public' courses, which do not require membership or a green card. In Finland none of the full-size golf courses are of this type.[11]

At the start of the twenty-first century, Finland had more than 100,000 golf club members. Golf was gaining strongly in popularity; a national survey of 2001–2002 showed that the number of people playing golf had the potential to double.[12] Also the demand for golf greatly exceeded the supply of courses, especially in the Helsinki metropolitan area.[13] In the city of Helsinki, there are only two full-size 18-hole courses, Tali golf course and Vuosaari golf course. The latter was built on a brownfield site in eastern Helsinki in 2002. The others are smaller and there are no new large golf courses planned in the city. However, one of the two 9-hole golf courses, an everyman's golf course, Paloheinä, was permitted an expansion of 15 hectares after a long debate in 2007. The nearby residents were against the expansion, as they felt the traditional landscape of agricultural fields was threatened. The fact that the Paloheinä golf course was to be open for everyone seems to have been the decisive factor in getting council approval for the expansion.[14] As the city of Helsinki lacks vacant areas for new golf course developments, most future courses will be located in other parts of the metropolitan area. The city of Espoo, west of Helsinki, has the largest amount of golf courses in the capital area (see Figure 9.1).[15]

Tali golf course: early years

The Tali golf course has been developed around the old manor of Tali, the history of which goes back to the early seventeenth century.[16] When the golf course was built in 1932, the area was situated relatively far away from the main residential areas. As the metropolitan area of Helsinki has grown, the relative location of the Tali course has changed. Today four residential areas surround the golf course: Pajamäki, Munkkivuori, Talinlehto and

---

11 'Miljonääri takaa uuden golfkentän'; The website for the Finnish Golf Union, <www.golf.fi>, Visited 25.3.2008; Tiainen, Golftoiminnan kansantaloudelliset vaikutukset.
12 Golfkenttäselvitys also listed the number of golf courses in Sweden (418), Denmark (135) and Norway (129). See Golfkenttäselvitys, 4–5.
13 Golfkenttäselvitys, 1; A. Manninen, 'Golfbuumi pyyhkii Espoota', *Helsingin Sanomat*, 24.9.2001.
14 Paloheinä Golf, Uutiset: Valtuusto päätti murskaäänin: Paloheinän kenttä laajenee!, <http://www.paloheinagolf.fi/>, Visited 26.1.2008.
15 Manninen, 'Golfbuumi pyyhkii Espoota'.
16 L. Kalliala, *HGK 1932–82*, Helsinki: Helsingin golfklubi 1982, 2.

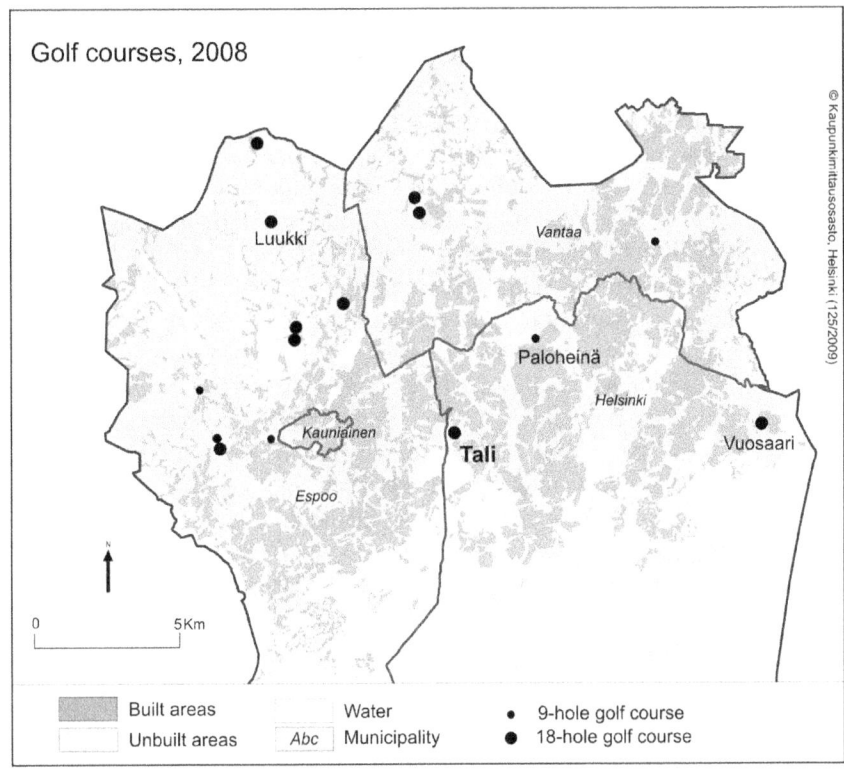

Figure 9.1. The Tali and other golf courses in the Helsinki capital area (Helsinki, Espoo and Vantaa), in 2008.[17] Golf courses mentioned in the text are indicated in the map.

Talinranta (Figure 9.2), built between the 1950s and the 1990s. Also, there is a sports park adjacent to the golf course. Unlike Tali, most of the newer golf courses in the Helsinki metropolitan area are situated on the urban fringe or in non-built up areas (Figure 9.1).

When the Tali golf course was established it had an area of 10 hectares, including the main building of the old Tali manor, fields and some forest. The area was not actually within the official city borders of Helsinki, but was part of the rural district. However, a rental agreement for golf course activity was established between the *Helsingin Golfklubi ry* and the city. The owner of Tali manor had sold the land to the city in 1930. At that time, the city of Helsinki was in the process of acquiring property rights outside its borders to ensure the needs of the growing capital. The golf course area in Tali, along with communities and rural areas around Helsinki, was merged with the city in the big annexation of 1946. During the first decades the

---

17 Basemap is from the digital map database SeutuCD'03 compiled by Pääkaupunkiseudun yhteistyövaltuuskunta YTV (The Helsinki Metropolitan Area Council). Data from golf courses is withdrawn from the website of Suomen Golfliitto (The Finnish Golf Union), <www.golf.fi>, Visited 22.9.2009.

contract was renewed every four to five years. From the beginning, the rented land was not meant exclusively for golf. One provision in the rental agreement permitted the city to grow grass on designated areas, and several other conditions have since been included in the agreement. For example, there was an obligation for the golf club to renovate the main building of Tali manor,[18] along with a provision that 'the leaseholder is obliged to give out parcels of land or buildings, if the city needs them to build housing, streets, or other infrastructure serving public interest...'.[19]

The planned Olympic Games at Helsinki in 1940 gave an impetus to build and develop a variety of modern sports facilities. Attention was given to golf, even though the game was not in the Olympic programme. In 1938, an agreement with the city allowed the expansion of the golf course up to 50 hectares. However, the Second World War interfered with the plans. The Olympics were cancelled, and the expansion of the golf course was postponed. During the Winter War 1939–1940, the main building of Tali manor was transformed for army use, and the nearby outhouses served as a prison. Part of the fairways was used for growing hay for the army's horses (most of the parks in Helsinki were used for military operations or as fields for horses during wartime). In 1942 the Agricultural Department of the city wanted to convert the entire golf course area into potato fields, but in the end only some roughs were set aside for the purpose. During the period only part of the golf course was maintained for playing.[20]

After the war years, the golf club rented more land from the city and extended the course area to the surrounding agricultural fields and meadows. In 1949 the golf club made its first 15-year rental agreement with the city. By 1952 the course area had been expanded to 39 hectares, and the golf course was able to establish a full-size (18-hole) programme. After renting a further 15 hectares of land from the city in 1964, the total length of the golf course measured up to international standards, and the rented area totalled 54 hectares.[21]

The golf course was not crowded with golfers at that time. Acquiring the club membership and playing there had become so expensive that only a few could afford it. To improve the possibility of playing golf in Helsinki, another golf club, *Suur-Helsingin Golfklubi ry (SHG)* was founded in 1965.

---

18 T. Manninen, *Helsingin Golfklubi 1932–2002*. Helsinki: Helsingin golfklubi 2002, 4, 18–20; Kalliala, *HGK 1932–82*, 42; Suomen urheiluarkisto (The Sports Archives of Finland), Helsingin Golfklubi, Vuokrasopimus Golfklubin ja kiinteistölautakunnan välillä 11.4.1932; Y. Harvia, *Helsingin esikaupunkiliitos*, Päämietintö (mietintö n:o 1). Helsinki: Helsingin kaupunki 1936–1940, 121, 140.

19 Suomen urheiluarkisto (The Sports Archives of Finland), Helsingin Golfklubi, Vuokrasopimus Golfklubin ja kiinteistölautakunnan välillä 1936.

20 Kalliala, *HGK 1932–82*, 38–46; T. Kopomaa, *Kaupunkipuistojen käytöt: Elämää Helsingin puistoissa ja ulkoilualueilla*. Helsingin kaupungin tietokeskuksen tutkimuksia 5. Helsinki: Helsingin kaupunki 1995, 18. Enhancing golfing facilities in the late 1930s was part of the general improvements of the cityscape and leisure activities for the Olympic visitors. Golf was on the Olympic programme in 1900 and 1904, but not afterwards: Suomen urheiluarkisto (The Sports Archives of Finland), Helsingin Golfklubi, Helsingin Golfiklubi r.y. – Helsingfors Golfklubb r.f. Årsberättelser 1939– 44.

21 Manninen, *Helsingin Golfklubi 1932–2002*, 18–20, 47.

Helsinki city rented the *SHG* a small field in the eastern Helsinki, in Rastila, but as the place proved too small to accommodate competitions, the golf club applied for a bigger area outside the city borders, in Luukki. Some city councillors advocated that this golf course should be a municipal one. They argued that this would make the golf course cheaper and more accessible for those wanting to play golf but unable to afford the high cost of membership at Tali. *SHG* received land in Luukki, owned by the city of Helsinki, but the golf course was not made a municipal one.[22] However, currently the Luukki golf course is open for players without a golf handicap, green card or a membership of a golf club for one day a week.[23]

Paradoxically, while the Tali golf club was renting more land for its own use in the years between 1949 and 1964, there started to be problems with finding public park space in Helsinki. At the end of the 1950s, the 'peoples parks' (*kansanpuistot*) near the central city areas, such as the parks on the Seurasaari and Pihlajasaari islands and those in Kivinokka and Tullisaari, suffered from serious overcrowding. To offer better recreation possibilities to its inhabitants, the city was active in the 1950s in buying large parcels of land outside its borders for recreation.[24] However, it seems there was no public discussion concerning the 54 hectares of park land reserved for the Tali golf club and its elite golfers. The lack of discussion at this time may be explained by the fact that the board of the Tali club consisted of leading Helsinki figures, who had, as will be shown, a close relationship with city politics.

At the start of the 1970s, however, the Tali golf course figured in growing public debate, and it became unclear whether the golf course would be continued in the area. The rental agreement was only renewed for five years from 1969, and discussions arose whether there would be a further agreement at all. One of the left-wing parties in the City Council was against renewing the rental agreement. *The Finnish People's Democratic League, SKDL* (Suomen Kansan Demokraattinen Liitto) argued in a City Council meeting in 1971 that [the renewal of the rental agreement:

> is unfair, when considering the following aspects... in Helsinki, there are a number of organisations, sports clubs and other citizen organisations that have thousands of members in one organisation, and regardless of applications have not been offered any suitable areas, or areas that have been very small...there is a lack of land for different public purposes, and in this sense, this matter should be reconsidered because the city has

---

22 L. Tilander, 'Suomalaisen golfin vuosikymmenet', in (ed.) *Suuri golfkirja 1*, Porvoo: WSOY 1996, 164–165; Helsingin kaupunginarkisto (Helsinki City Archives), Helsingin kaupunginvaltuuston asiakirjat, kaupunginvaltuuston päätökset, 2.11.1966, n:o 46. City of Helsinki had bought the area around Luukki for recreation in 1961.

23 Golfpiste.com, Suur-Helsingin Golf, Luukki ja Lakisto, <http://www.golfpiste.com/kentat/shg/?lang=fi>, Visited 2.1.2008. A 'golf handicap' is special golf terminology and refers to the numerical system, which allows golfers of different abilities to compete on the same level. Golf player's home club or golfing association issues handicap certificates that indicates his or hers current handicap.

24 K. Ilmanen, *Ensimmäisenä liikkeellä: Helsingin kaupungin liikuntatoimi 1919–1994*. Helsinki: Helsingin kaupungin liikuntavirasto 1994, 97.

bought thousands of hectares of land from the nearby municipalities, so would it not be possible to move this type of area to one of the nearby municipalities.[25]

The City Planning Department had also become interested in the area. As part of the preparations for the Helsinki Master Plan 1972, there had been studies about the possibility of different uses for the golf course. One of the planning documents of 1971 for Tali–Iso-Huopalahti stated:

> The most significant goal for the Iso-Huopalahti–Tali area is to reserve and further develop it as a recreational area...; currently the area is mainly serving specific groups [football players and golfers], or those who are able to buy patches from the Tali allotment. The area is also used for outdoor recreation and skiing, but the municipal tip, Tali water treatment plant and the rough wasteland areas encumber these activities. There is a lack of areas for active recreation and sports for the larger public. [26]

The municipal tip and the water treatment plant (see Figure 9.2) were built in 1963 and 1956 respectively. Generally, the proposal for the Helsinki Master Plan 1972 advised that recreation functions serving a relatively small proportion of population and requiring large areas should be located outside the city borders.[27] The City Planning Department also had plans to expand westward one of the nearby residential areas, Pajamäki. This would affect some of the fairways at the northern part of the golf course. In addition, there were schemes for building a summer theatre, a disco and a restaurant in the Tali manor area.[28]

The sudden enthusiasm for converting parts of the Tali golf course to other uses can be seen as part of the change in city planning at this time in Helsinki. The development of city planning policy in Helsinki can be roughly divided into two phases, with the turning point marked by the 1972 Master Plan. The earlier period was characterised by policies to accommodate the population growth of Helsinki in dispersed suburbs (often characterised by high apartment buildings). After 1972 planning policy sought to solve problems of urban growth by promoting more compact housing areas and limiting urban dispersal. In addition to the change in the focus of planning, the sheer magnitude of planning activities grew significantly. In Finland rapid urbanisation and increasing municipal obligations contributed to a breakthrough in community planning after 1965. The demand for planning was especially strong in the metropolitan area. Helsinki's City Planning Department was established in 1964 and the number of planners significantly increased in the following decade. Also about this time, the idea of 'planning

---

25 Helsingin kaupunginarkisto (Helsinki City Archives), Helsingin kaupunginvaltuuston pöytäkirja 23.6.1971, 435 §, esityslistan asia n:o 10.
26 *Iso-Huopalahti yleissuunnitelma: Perusselvitysraportti 1971*. Helsinki: Oy Kaupunkisuunnittelu Ab 1971, 31, 35.
27 Helsingin Kaupunkisuunnitteluvirasto, Yleiskaavaosasto, Helsingin yleiskaava 72: Virkistys: Esiselvitykset 30.9.1971.
28 'Helsinki ja Espoo yhteistyössä: Iso-Huopalahden-Talin alue kehitetään virkistyskäyttöön', *Helsingin Sanomat* 8.10.1971.

democracy' emerged, and different citizen groups and stakeholders started to raise their voices on planning issues.[29]

Against this background, it is not surprising that the large green space at Tali serving a private golf course figured in planning discussions. Criticism also came from other directions. After the 1960s public funding for sports sites started to increase, and the city built a number of new sporting sites, such as football grounds next to schools, recreation routes in forests and sports parks accommodating multiple forms of sport. The allocation of space for different sports was actively discussed in newspapers and the Tali golf course came under fire. In 1973 an article in the main daily newspaper, *Helsingin Sanomat,* argued that the amount of maintained green space for golfers was disproportionate. It was much more than that reserved for football players, who were far more numerous than golfers.[30] Again, the Social-Democratic newspaper, *Suomen Sosiaalidemokraatti*, compared the Tali golf course to the multi-functional Pirkkola sports park opened in 1968. It reminded readers that the sports park had a size similar to the golf course, but offered sports and recreation services for a much greater number of people.[31] Ultimately, one of the three biggest sports parks planned in Helsinki was built for Tali, next to the golf course area.[32]

Despite all the discussion over the Tali golf course, the city of Helsinki made a new rental agreement with *Helsingin Golfklubi* in 1971. The new agreement was made for 15 years. It seems clear that, regardless of all the plans and debate, the golf club wielded some influence over the final decision. In the 1970s golf remained popular among city politicians, who had the power to save the golf course. To take one example, the deputy chairman of the Board of *Helsingin Golfklubi*, Gunnar Smeds, was also the deputy mayor of the city of Helsinki at the time. According to the official history of *Helsingin Golfklubi*, his impact on the outcome was influential.[33]

The golf course from the 1980s

In 1984, the rental agreement was renewed for 25 more years until the end of 2009. In the request for the renewal, the golf club gave, more or less, the same reasons as 15 years earlier:

29 L. Kolbe, 'Helsinki kasvaa suurkaupungiksi: Julkisuus, politiikka, hallinto ja kansalaiset 1945–2000', in L. Kolbe and H. Helin, *Helsingin historia vuodesta 1945:3*. Helsinki: Helsingin kaupunki 2002, 284–286; T. Herranen, 'Kaupunkisuunnittelu ja asuminen', in O. Turpeinen, T. Herranen and K. Hoffman, *Helsingin historia vuodesta 1945:1*. Helsinki: Helsingin kaupunki 1997, 215–217; H. Schulman, 'Helsingin suunnittelu ja rakentuminen' in H. Schulman, P. Pulma and S. Aalto, *Helsingin historia vuodesta 1945:2*. Helsinki: Helsingin kaupunki, 2000, 46, 51; M. Kirjakka, *Yhdyskuntasuunnittelun henkilömäärien kehitys 1970- ja 1980-luvuilla*, Espoo: Yhdyskuntasuunnittelun täydennyskoulutuskeskus 1988, 10, 27.
30 'Itä-Helsinki kokonaan paitsiossa: Nurmikenttiä saatava tarpeeksi eri puolille', *Helsingin Sanomat* 10.5.1973.
31 'Golfpallolla päähän Talissa laillisesti', *Suomen sosiaalidemokraatti*, 2.7.1968.
32 'Kuusi urheiluhallia luonnonläheisyyteen: Talin puisto hahmottumassa', *Uusi Suomi*, 25.10.1974.
33 Manninen, *Helsingin Golfklubi 1932–2002*, 30.

the importance of the golf course as the only one in Finland meeting international standards, the need for long-term renovation planning, the suitability of the sports for people of all ages, and the fact that many large cities in the world have golf courses in their vicinity.

No differing opinions were documented in the City Council files. The Sports and Recreation Board of the city supported the renewal of the contract until 2009, though in 1971 the Board had regarded 25 years as too long for a contract.[34]

In Helsinki, the pressure to develop more land for housing had calmed down by the 1980s, when the city's projected population growth was not realised.[35] However, the urban structure was getting denser around the Tali golf course, as a new residential area, Talinranta, south of the golf course, replaced the old water treatment plant in the late 1980s. One of the planned streets leading to Talinranta crossed a course fairway. As compensation, part of the eastern border of the golf club was moved eastward to the Tali sports park area, and two hectares of the sports park fields were given to the golf club. It seems the compensation was regarded as fair on both sides as the exchange of land areas did not result any conflict. After these changes in the late 1980s, the borders of the course stayed the same.[36]

The economic recession in Finland in the early 1990s did not seem to have much effect on the functioning of the Tali golf course, unlike the case with the public parks of Helsinki which were run down as a consequence of financial cutbacks. *Helsingin Golfklubi* had healthy finances, and renovations to some of the Tali manor buildings were done, with extra funding raised by the club's members.[37] The protected status of Tali manor was enhanced at this time. The Helsinki Master Plan 1992 recognised it as a site with cultural-historical significance, and the following year the buildings at Tali manor were put under the protection of the National Board of Antiquities, which is responsible for Finland's built cultural heritage.[38]

Zoning through town plans can offer sports and recreation areas greater stability, and better legal standing against other land use development.[39]

---

34 Helsingin kaupunginarkisto (Helsinki City Archives), Helsingin kaupunginvaltuuston asiakirjat, kaupunginvaltuuston päätökset, 8.6.1983, n:o 10; 23.6.1971, n:o 10.
35 L. Kolbe, 'Helsinki kasvaa suurkaupungiksi', 431.
36 Manninen, *Helsingin Golfklubi 1932–2002*, 38; T. Manninen, *HGK: Muutosten vuodet 1982–1992*, Helsinki: Helsingin Golfklubi 1992, 9–11; Helsingin kaupunginarkisto (Helsinki City Archives), Helsingin kaupunginvaltuuston asiakirjat, kaupunginvaltuuston päätökset, 8.2.1989, n:o 3.
37 The funding for Park Division of the Public Works Department of Helsinki, which maintains most of the public parks (forested areas, meadows, grass lawns, children's playgrounds with small ballgame grounds) in Helsinki, was continuously diminished from 1990 to 1997, which caused dilapidation of the parks. Helsingin rakennusviraston arkisto (The City of Helsinki Public Works Department Archives), Helsingin rakennusviraston toimintakertomukset 1990–1997; Suomen urheiluarkisto (The Sports Archives of Finland), Helsingin Golfklubi, *HGK:n Uutiset* –lehti 1991–2000; Manninen, *Helsingin Golfklubi 1932–2002*, 38–42.
38 Helsingin kaupunkisuunnitteluvirasto, Talin kartanon alue, asemakaavan ja asemakaavan muutoksen nro 11120, selostus (Päivätty 29.8.2002), 5.
39 V. Rajaniemi, *Liikuntapaikkarakentaminen ja maankäytön suunnittelu: Tutkimus eri väestöryhmät tasapuolisesti huomioon ottavasta liikuntapaikkasuunnittelusta ja sen kytkemisestä*

Before 2002, the Tali golf course area had been designated for recreation purposes in Master Plans, but no detailed town plan concerning the area was in effect. The city of Helsinki decided to make a town plan for the Tali area before the next renewal of the rental agreement in 2009. With the town plan the golf course would have a legal basis as long as the town plan itself was not changed. Consequently, a town plan for Tali manor with the surrounding golf course was included in the new Helsinki Master Plan in 2002. In addition to enhancing the significance of Tali manor as a cultural-historical site, the golf course itself was recognised as 'a sports and recreation area with cultural-historical and landscape significance, which should be maintained and renovated in such a manner that its characteristics will be preserved.'[40]

Tali golf club and the surrounding community

At the beginning, in 1932, the Tali golf club had 115 members. The number has increased steadily throughout the life of the golf club, except for the time immediately after the war, when there were only 30 active players, and the number of passive members had dropped from over 200 to 60.[41] In the late 1970s the membership numbers kept rising and started to reach the then membership limit, 800. Guest players were required to have membership of a Finnish golf club and a golf handicap. In 1979 the crowding of the golf course was alleviated by setting restrictions on usage, which were extended to the guest players in 1988. At the present time, the course is very busy in the playing season. Since the latest expansion of the golf course area in 1964, the membership has more than doubled to 1,450 members (in 2005), as shown in Table 9.1.

The average Finnish golfer is said to be male aged between 35 and 50. However, the proportion of women has grown markedly. In 2005, female membership was 35 per cent at Tali, against an average of 28 per cent for all the clubs in Finland. From the 1980s golf was gaining more status as a leisure activity for the whole family. The proportion of junior players in Finnish golf clubs has fluctuated between 11 and 15 per cent in the years 1960 to 2005. At Tali, the figure doubled after 1982, and currently stands at 19 per cent. The biggest growth occurred around the turn the century.[42] The increased number of junior players may be explained by a policy introduced in the 1990s. Because of the concern over the aging membership profile of

---

maankäyttö- ja rakennuslain mukaiseen kaavoitukseen, Studies in sport, physical education and health 109. Jyväskylä: Jyväskylän yliopisto 2005, 87.

40 The quote is from the town plan: Talin kartanon alue, 8; Golfliitto, Uutisarkisto, 'Talin golfkentän alue suojelukohteeksi', 21.3.2003. <http://www.golf.fi/portal/uutiset/piccadilly/?bid=2526&vid=12>Visited 21.9.2006; L. Lukkarinen-Annila, Architect, Helsinki Planning Department. Personal information in an interview 11.5.2006.

41 Kalliala, *HGK 1932–82*, 46.

42 Kalliala, *HGK 1932–82*, 114; H. Heininen, office secretary, Helsingin Golfklubi ry, personal communication by email, 26.9.2006; O. Saarinen, Head of the Organisation, Suomen Golf-liitto (The Finnish Golf Union), personal communication by email, 31.8.2006.

the club, the younger generation, and especially the offspring of the member families, were encouraged to join the club.[43] Today one acquires a membership to the Tali golf course only by queuing for a very long time – 10–15 years. Also, a reference from two club members is required. However, in the latest renewal agreement of the rental agreement for the golf course at the end of 2008, the city increased significantly the rental fees and, for instance, ordered more playing hours for non-members.[44]

Table 9.1 Number of members in *Helsingin Golfklubi ry*, and all Finnish golf clubs, 1970–2005.

Year	Golf clubs in Finland Members	Helsingin Golfklubi ry. Members
1970	1,776	625
1975	2,145	825
1980	3,211	891
1985	6,671	1,051
1990	29,934	1,225
1995	45,272	1,246
2000	76,522	1,371
2005	110,156	1,450

Given that golf was played by only a small minority for a long time, it understandably had the reputation of being an elitist game. Thus, it is interesting to examine in more detail the changing social status of the membership at the *Helsingin Golfklubi*. The period up to the 1960s has been claimed as the most select period in the club's history. The *Helsingin Golfklubi* was founded by five men, well known in business and politics: two foreigners, the Dutch businessman Charles Jensen and the US ambassador to Helsinki Edward Brodie, along with the bank manager Eero Ilves, Erik von Frenckell, the deputy mayor of Helsinki (1931–1955), and Eljas Erkko, who acted as the chairman of the club board between 1947 and 1960. Erkko was a minister and chief-editor (1931–1938) of the biggest newspaper in Finland, *Helsingin Sanomat*. He is said to have personified the emerging golf culture in Finland in the 1950s. In those days, golf was said to be 'Erkko's game'. The club board consisted largely of the same men for the first decades of the club's history.[45]

The club's early years had the atmosphere of a country club with the social gatherings and prestige dinner parties at the club house being nearly as important as playing the game itself. The most spoken languages at the

---

43 Suomen urheiluarkisto (The Sports Archives of Finland), Helsingin Golfklubi, J. Schauman, puheenjohtajan palsta, *HGK:n Uutiset* –lehti, helmikuu 1993.
44 Website for Helsingin Golfklubi ry. <http://golfpiste.com/kentat/hgk/jasenasiat/?p=jasenyys &lang=fiR> Visited 15.8.2009.; Jokinen & S. Laita, 'Talin kentän vuokrakiistassa häämöttää vihdoin sopu', *Helsingin Sanomat* 19.11.2008.
45 Kalliala, *HGK 1932–82*, 16–18, 34; H. Brotherus, *Eljas Erkko, legenda jo eläessään*, Porvoo: WSOY 1973, 156–160.

club were Swedish and English; it was only in 1956 that the club rules were translated into Finnish. Nor did the social composition of the golf club membership change much in subsequent decades. According to the 1980 yearbook of the club, roughly 80 per cent of adult male members had higher education or a profession requiring higher education. The remainder consisted of students (13 per cent), and those with lower education (7 per cent).[46] The year 1980 is the last year when the occupations or status of members were given. Thus, it is not possible to see if the membership profile changed during the golf boom of the 1980s. Certainly, the club's activity became more centred on the game itself. The 'vanished atmosphere of a country club' was discussed in the *HGK* newsletter in the 1990s.[47] The crowding of the golf course gradually resulted in stricter and more efficient rules for playing times and the players' performance at the course. Playing golf became a sport with no extra luxuries.

What about the impact of the course on the local community? Potentially, the location of a golf course, equivalent to a large private lot in an increasingly densely built up urban area with relatively little other green space, has the risk of causing friction with local residents. Here the multifunctionality of the golf course is a key issue. In Tali's case, the city of Helsinki has sought to ensure the multiple use of the area, to reduce this risk. As discussed above, the rental agreement had terms of use already from the 1930s. To some extent, the Tali golf course has always been open to walkers, as well as to joggers and other types of recreation. Only areas dangerous to the passers-by are fenced, or have warning signs during the playing season.[48] As well as the open-access policy in some parts of the Tali golf course, the city has also sought to increase the number of official paths crossing the course.[49] In Helsinki, there is a network of official recreational routes, which connect the recreational green space to residential areas. In 1979, the official recreation routes consisted of one route through the golf course and Tali manor, and one on the eastern side of the course. Recreation opportunities around the golf course were improved in the 1980s when the waste tip (see Figure 9.2) was closed and landscaped for recreation.[50] Also, when the border of the golf course was moved eastward in 1989 (see above), the recreation route on the eastern side of the course was also moved.[51]

46 Kalliala, *HGK 1932–82*, 34. The data on titles and professions of the male club members were gathered from the yearbook of the Helsingin Golfklubi in 1980. The year 1980 was the last year when membership records show profession or education data on members: Suomen urheiluarkisto (The Sports Archives of Finland), Helsingin Golfklubi, Helsingin Golfklubi – Helsingfors Golfklubb 1980, Helsinki 1980.
47 Suomen urheiluarkisto (The Sports Archives of Finland), Helsingin Golfklubi, J. Schauman, puheenjohtajan palsta, *HGK:n Uutiset* -lehti, huhtikuu 1992.
48 Manninen, *HGK: Muutosten vuodet 1982–1992*, 18; Lukkarinen-Annila, 11.5.2006.
49 Lukkarinen-Annila, personal information via telephone 21.9.2006.
50 The municipal tip west of the golf course was in use between 1963–1979: see *Tutkimus Iso-Huopalahden suljetun kaatopaikan ympäristöhaitoista*. Helsingin kaupungin ympäristönsuojelulautakunnan julkaisuja 4/1988. Helsinki: Helsingin kaupunginkanslia 1988; *Ulkoilureittisuunnitelma, osa II: Pääulkoilureitistö perusteluineen*. Helsingin kaupungin kaupunkisuunnitteluviraston yleiskaavaosaston julkaisuja YB 4/1982.
51 Helsingin kaupunginarkisto (Helsinki City Archives), Helsingin kaupunginvaltuuston asiakirjat, kaupunginvaltuuston päätökset, 8.2.1989, n:o 3.

Seasonality in Finland also offers considerable opportunities for the multifunctional use of the golf course. In Tali, skiing is allowed on the golf course area in winter on designated tracks, which are maintained by the city of Helsinki Sports Department. The official track route in Tali was specified in a planning document in 1978, and the route remains more or less the same at the present time.[52]

The 2002 town plan for the Tali area included a proposal to increase the number of public recreation paths in the golf course area. The Planning Department holds the view that residents living in the neighbourhood are not against the golf course as long as their recreational opportunities are assured. In fact, in the participatory process for the Master Plan 2002, comments from the public showed that the golf course was regarded as important in protecting the area's recreation and environment from future housing development.[53] Not least important, it is likely that the presence of the course may well increase the property value of housing in the vicinity.

Major disputes between the golf club and the surrounding residents were not visible in the public documents used in this study. Only one case was found where a neighbourhood association from the residential area north of the golf course, *Pajamäki seura ry*, tried to appeal to the city in 1993 to stop the building of a four meter high shield fence to the driving range in the south part of the course. The association argued that fencing that area would significantly affect the scenery and hinder the recreation possibilities at the golf course outside the golfing season. The appeal was rejected, because the golf club justified the building of the fence as protecting for the people on the jogging routes. The new residential area, Talinranta, south of the golf course had made it necessary to move the 'driving' direction on the range.[54] Generally the course is increasingly seen as a social and cultural asset for the area.

Environmental implications: golf course maintenance

The city became interested in the social implications of the Tali golf course area at the start of the 1970s. At that time, the golf course and its surrounding areas also received environmental study. For the Master Plan 1972 the city contracted out studies of the recreation areas. In the case of the Tali golf course and the surrounding green space, environmental features, such as resistance to erosion and the reproduction of animal and plant populations, were evaluated. Among the recommendations, the study called for the protection of a strip of groves along the stream Mätäjoki running through

---

52 *Ulkoilureittisuunnitelma: osa 1: Perusselvitykset ja tavoitteet*, Helsingin kaupungin kaupunkisuunnitteluviraston yleiskaavaosaston julkaisuja, YB 21/1978, Liite 7. Hiihtoreitit ja talviulkoilupaikat 1.3.1979; T. Marjamäki, field manager, Helsingin Golfklubi, personal communication in an interview 23.10.2006.
53 Lukkarinen-Annila, 11.5.2006; Talin kartanon alue, Liite 9, Ulkoilureitistö; M. Mattila (ed.), *Helsingin yleiskaava 2002, Osallistuminen ja arviointi, Asukkaiden yleiskaava*. Helsingin kaupunkisuunnitteluviraston kaavoitusosaston selvityksiä 1/2000, 55.
54 Liikuntaviraston arkisto (The City of Helsinki Sports Department Archives), Helsingin kaupungin rakennuslautakunnan päätös 15.6.1993, 468 §.

Land use change around the Tali golf course

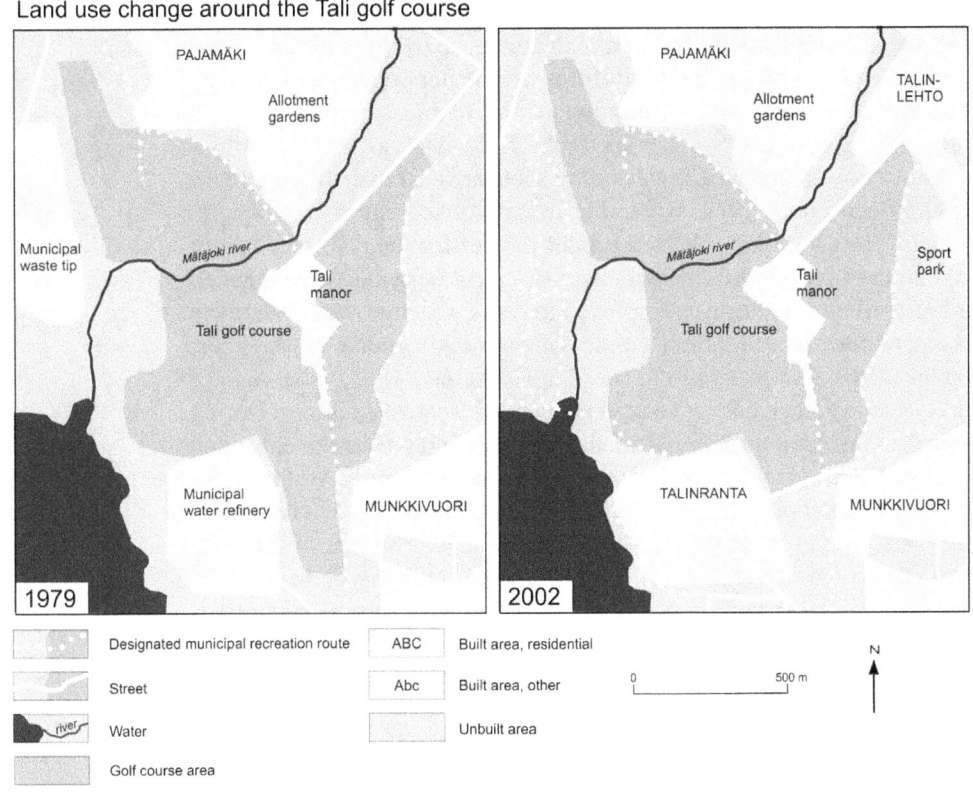

Figure 9.2 Land use change around the Tali golf course, 1979–2002[55]

the golf course.[56] Before that, there was only one protected site at the golf course: the largest oak in Helsinki was given protection in 1958. Generally, the late 1960s was a period when environmental questions started to attract attention in Helsinki. In 1971 the city established the first special environmental protection committee, which was to become the Environment Board (*ympäristölautakunta*) in 1984.[57] Still today, the woodlands around the Mätäjoki are protected. In 2008, the golf club even claimed that the conservation of the area limits 'effective course management'.[58]

---

55 The maps are redrawn by hand from several sources: Ulkoilureittisuunnitelma: Osa 1, Liite 4. Ulkoilutiet hallintokunnittain 31.5.1979.; Talin kartanon alue, liite 9, ulkoilureitistö.; Kiinteistökartta 1969, 2003, Kaupunginmittausosasto, Helsingin kaupunki.; SeutuCD'03.
56 *Iso-Huopalahti yleissuunnitelma*: *Perusselvitysraportti 1971*, 15.
57 Kolbe, 'Helsinki kasvaa suurkaupungiksi', 358.
58 The statement is one of arguments the golf club used to plead for reduced rental fees in the renewal process of the rental agreement in 2008. See the arguments recorded in the Helsinki City Board Agenda 1.12.2008 (Khs 2008-2462. Talin golfkentän maa-aluetta koskeva vuokrasopimus Helsingin Golfklubi Ry:n kanssa.) at the website Golfpiste < www.golfpiste.com/kentat/hgk/upload/Vuokrasopimusehdotus.pdf>. Visited 15.8.2009.

How has the golf course area been maintained over time and how has this affected the environment? As well as the course being expanded and new playing areas having to be built, maintenance operations at Tali have experienced changes including a shift from rudimentary operations done by hand to a highly mechanised trimming of the teeing areas, greens and fairways. In the mid-1940s the grass at the fairways was still cut with a horse-drawn mower, and the fairways were a lot narrower than today. Greens were cut by hand until 1954, when the first motorised cutter was bought. The roughs grew long grass until a proper mower for them was bought at the beginning of the 1950s. The amount of short-cut fairways was expanded in the 1950s when a tractor was employed to pull a wider mower. Due to the increasing number of golfers and a more congested golf course in the 1970s, the roughs were cut even shorter to speed up the game.[59] Today, the ratio of roughs, fairways and greens fluctuates between different golf seasons. During the international competitions at Tali, the fairways and greens are expanded to meet the requirement for a smoother golf game. Currently, of the entire 54 hectares of the course area in Tali, there is one hectare of greens, 1.5 hectares of teeing areas and 10 hectares of fairways. The rest of the area, 41.5 hectares, consists of roughs, bushes, woods, bunkers or passages.[60] The golf course has a very park-like appearance, as there are many trees between the fairways.[61]

The Tali golf course has suffered more from problems of wetness than of drought. The city of Helsinki dredged the lower reaches of the Mätäjoki stream in the late 1940s, alleviating flooding in the spring and in the autumn. There was also some subsurface drainage done in the 1930s, but the main maintenance operations were inaugurated in the 1980s. The new 25-year lease contract in 1984 enabled long-term plans for the course, and thus a systematic renewal of the course was carried out. The current playing areas emerged then. By the mid-1980s subsurface drainage was completed on the fairways that suffered during rainy seasons; several kilometres of drainage were dug. Between 1998 and 2002 more subsurface draining was completed.[62] As well as drainage, irrigation was needed at the course, and a motorised pump was already bought for this purpose in the 1930s, though a computer-aided irrigation system was not completed until the mid-1980s. At that time, the golf club dug a bigger irrigation pool from the Mätäjoki stream and built a small dam to it.[63] Currently, the part of Mätäjoki within the golf course area is dredged once a year. Using the Mätäjoki stream for irrigation has an impact on the lower reaches of the stream. During the dry season in the summer the stream flow decreases due to irrigation at the golf course. Pumping the water to the golf course is allowed, as it is not considered to

---

59 Kalliala, *HGK 1932–82*, 38, 106; A. Arkkola. A long-time member of the Helsingin Golfklubi ry and a member of a field committee of the golf club. Personal communication by emails 30.10.2006 and 1.11.2006.
60 Marjamäki, 23.10.2006.
61 Manninen, *HGK: Muutosten vuodet 1982–1992*, 8.
62 Manninen, *Helsingin Golfklubi 1932–2002*, 4–5; Manninen, *HGK: Muutosten vuodet 1982–1992*, 4–8; Marjamäki, 23.10.2006.
63 Kalliala, *HGK 1932–82*, 38; Manninen, *HGK: Muutosten vuodet 1982–1992*, 3–4.

have a severe effect on the Mätäjoki stream. However, the whole waterway of Mätäjoki often seems to suffer from some dryness during the summer months. At that time, the cities of Helsinki and Vantaa buy more water from the Helsinki Water company (Helsingin Vesi), a municipal utility, to pump into the upper reaches of the Mätäjoki stream to enhance the water quality and maintain the scenic value of the river in the recreation areas.[64]

In addition to being used for irrigation purposes, the Mätäjoki stream may also be affected by other maintenance operations at the golf course. Concerns about fertilisation and pesticide residuals and their impact on the surface water from intensively managed golf course greens and fairways arose for the first time in the 1980s. In their defence, golf clubs usually claim that they do not use excessive amounts of fertilizers, not least because this results in the grass growing faster and requiring more frequent mowing.[65] In Finland, due to the relatively cold climate, there are not as many pests and weeds as in warmer climates, and this may lead to more subtle maintenance practises in terms of pesticide usage.[66] In the Mätäjoki case, there is no research data available on the possible effect of the golf course. It is assumed, that the old waste dumping ground west of the course, now landscaped, and the former Tali water treatment plant (Figure 9.2) have a bigger effect on the Mätäjoki stream than the golf course.[67]

It is likely that the 1980s was the peak time for the use of fertilizers and pesticides at the Tali golf course. Unfortunately, there are no series of documents available that could be used to describe the changes in the type and amount of products used, but a little light can be shed. In 1936, only 1,500 Finnish markkas were spent on fertilizers out of a total club budget of 54,300 markkas.[68] According to one of the former field managers and a member of the golf club field committee Asko Arkkola, after the World War II fertilizers and pesticides were used whenever they were available, but a lack of money hindered usege until the 1980s, when improved club finances and the greater availability of commercial products facilitated their more extensive use.[69] At the present, long-term fertilizers are used at Tali,

---

64 Marjamäki, 23.10.2006; O. Ruth, university lecturer, Laboratory of Physical Geography, University of Helsinki, Personal communication by emails 26.10.2006 and 2.11.2006; P. Niemi, environmental inspector, city of Helsinki, Environment Centre. Personal communication by email 2.10.2006. Upper-reaches of the waterway within the borders of the city of Vantaa are called Mätäoja. Environmental history of Mätäjoki is also discussed in O. Ruth and M. Tikkanen, 'Purojen Helsinki, virtaava vesi kaupungin kahleissa' in S. Laakkonen, S. Laurila, P. Kansanen and H. Schulman (eds), *Näkökulmia Helsingin ympäristöhistoriaan: Kaupungin ja ympäristön muutos 1800- ja 1900-luvuilla*. Helsinki: Helsingin kaupungin tietokeskus, Edita 2001, 167–172.
65 T. Tynjälä, 'Nurmon Ruuhikoskelle tuorein ympäristöohjelma', *Etelä-Suomen Sanomat* 20.11.1998.
66 Suoninen, Porttikivi, Särkioja, and Taipalinen, *Tarinaharjun golfkentän pinta- ja pohjavesivaikutukset*, 19.
67 O. Ruth, 26.10.2006.
68 Suomen urheiluarkisto (The Sports Archives of Finland); Helsingin Golfklubi, F2 Kirjeistö 1932–1963, Golfkenttätöiden kustannusarvio vuodelle 1936 (Fairwayn katkaisu käsikoneilla). Salaries (31 000 markkas) composed the major part of the budget.
69 Arkkola, 30.10.2006 and 1.11.2006.

which is also the common policy in public parks in Helsinki.[70] Earlier, fertilizers used at the Tali golf course were the same as those used in agriculture, but they tended to induce the grass to grow too quickly, and thus complicated the mowing.

According to the field managers who worked at the golf course in the 1980s, Asko Arkkola and Timo Haapanen, the greens were treated mechanically in those days, and teeing grounds, fairways and roughs were treated chemically. Due to the lack of money this was done once in 1–5 years. Plantains, clovers and dandelions were controlled with pesticides. According to the current field manager, Teija Marjamäki, nowadays pesticides are only used to control plantains, once a year in the summer. Greens are still treated mechanically, and other playing areas chemically. Currently, the use of pesticides is limited to a very few products compared to those used in the 1980s.[71]

Conclusion

The oldest golf course in Finland, with its long history enriched by famous club members and the historic setting around Tali manor, offers an interesting study case. *Helsingin Golfklubi* established the golf course in the middle of agricultural fields, relatively far away from the main residential areas of the city. In the course of time, the city with its densely built suburban areas has surrounded the golf course. During the early 1970s, at the height of the planning movement in Helsinki, questions were raised about the future of the golf course, and whether it occupied too large an area in the growing city. Subsequent city plans, however, did not propose any alternative land uses for the golf course. On the contrary, the legal standing of the golf course was enhanced, the latest development being the recognition of the golf course area as a 'cultural milieu'. Whether this is due to a real desire of the city authorities to give the area such recognition, or more to effective lobbying by the golf club, remains unclear, and offers avenues for further research. However, the ownership set-up for the golf course, where the city owns the land and the golf club rents it, has inevitably created a mutual interest in negotiation and compromise. In all likelihood, the possibilities for the public to have access to the area for recreation would not have been as good, if the land had been under private ownership. On the environmental front, crucial questions remain whether the golf course is able to maintain a rich bio-diversity, and whether the environmental regulations are enough to mitigate the impacts from maintenance practises. Irrespective of the answers to these questions, continuing pressure for new housing land in Helsinki will keep the golf course on the city agenda for the foreseeable future.

---

70 H. Värri. Head of the Office, Street and Park Division, Helsinki City Public Works Department, personal communication in an interview 16.5.2006.
71 Arkkola, 30.10.2006 and 1.11.2006; Marjamäki, 23.10.2006; T. Haapanen, the field manager at the Tali golf course 1986–1991, personal communication by email 5.11.2006.

PHILIP JAMES AND EMMA L. GARDNER

10 The Role of Privately Owned Sports Related Green Spaces in Urban Ecological Frameworks

An ecological framework seeks to maintain ecological processes in the wider landscape and to conserve ecosystems, habitats, species, genetic diversity, and landscapes of importance. As greater attention is paid to ecological frameworks and in particular to such frameworks within an urban setting, then an understanding of the landscape ecology of sports related open spaces and their position within the wider ecological setting of a city requires attention. In this chapter we focus on golf courses and in particular address questions relating to their historic development and their contemporary role in urban ecosystems. The exploration of these issues will be based on a case study centred on a new golf course development constructed in the 1990s at the Marriott Worsley Park, Salford UK. In this case study the historic development of Salford is outlined and the development of the Marriott Worsley Park is discussed in detail. Contemporary land use data are presented in order to understand the spatial importance of sports and open space within the city. This analysis leads to suggestions for the inclusion of sports space, and in particular golf courses within urban ecological frameworks.

Land use changes in Salford

Salford is a historic settlement, noted in the Anglo-Saxon Chronicle and the Doomsday Book.[1] The name Salford originates from *sal*, the Latin *Salix* meaning sallow or willow, and 'ford', the only place to cross the River Irwell for many miles in either direction. It was the willow tree-lined banks of the Irwell that separated Salford from the, now better known, city of Manchester. Salford developed in the later eighteenth and nineteenth centuries from a village on the banks of the Irwell, to become an industrial city. Its industry was based on cloth, silk weaving, dyeing and bleaching.[2] The proximity of the Manchester Ship Canal, and Salford's own docks,

---

1 P. Teague, *The Salford Eye – A Background*, Salford Eye. 2000 Available internet: http://www.geocities.com/Paulontheair/salf3.html. Accessed on: 19 March 2002.
2 *Salford Local History*, Salford: Salford City Council 2003.

meant that the city was at the hub of the transport infrastructure, procuring raw materials and distributing manufactured goods from the Northwest.

Poverty and industrial squalor were dominant features of life in the Victorian city. Houses built for the growing population of industrial and manufacturing workers were of a poor standard; these cramped houses, as many as 200 per hectare, with poor sanitation, were crowded into 'slums'. During the late 1950s the slums were destroyed. People were moved to new housing estates, built throughout suburban and rural Salford. For example, it was at this time that the village of Little Hulton, 'Little Hill Town', was transformed from an idyllic village into a mosaic of council housing estates. Four large house-building phases occurred throughout the 1950s and 1960s, converting agricultural land into houses, shops, schools and roads. The area's natural environment and community characteristics were lost to over-development, social decay and degradation. In this way the scars from Salford's industrial past were extended both in time and space.

At the start of the twenty-first century Salford is a dynamic city, which has undergone radical transformations since becoming one of the world's first industrial cities.[3] Through extensive development and regeneration the city has been transformed from industrial decline and obscurity into a cultural and modern city. For example, Salford docks closed in the 1970s and the surrounding area spiralled into disuse and neglect accompanied by associated economic and social issues. During 1987, the Trafford Development Corporation was set up to encourage and attract new businesses into the region. This led to the transformation of Salford docks into Salford Quays, a hub of service industries, luxury water-front homes, and cultural and entertainment facilities.

Since the late 1980s the city has been constantly renewing itself and tackling the social, economic and physical implications of industrial decline.[4] The city of Salford has set out to reverse decay and degradation through a range of initiatives establishing itself with an international reputation for successful regeneration.[5] Today, largely as a result of regeneration activities, Salford is a city of contrasts: thriving business districts of Central Salford and Salford Quays, extensive inner city areas of Ordsall, Weaste, Eccles, Pendleton, Broughton and Blackfriars, and the picturesque village of Worsley. These built environments give way to large tracts of open countryside, much of which is prime agricultural land created from the former peat rich areas (known locally as mosslands) of Chat Moss, Linnyshaw Moss and Clifton Moss. Much of this agricultural land (approximately one third of the total area of Salford) is designated as Green Belt. The fundamental aim of Green Belt policy is to prevent urban sprawl by keeping land permanently open.[6]

---

3 City of Salford, *Salford Partnership. 1998. Building Sustainable Communities. A Regeneration Strategy for Salford,* Salford: City of Salford 1998.
4 City of Salford, *Milestones: The Story of Salford Quays*, Salford: Salford City Council 2005.
5 City of Salford, *Strategy and Regeneration Salford*, Salford: Salford City Council 2002.
6 City of Salford, *Salford Partnership.1998.*

Land use data compiled by the UK Government show that 55.7 per cent of the land in Salford is classified as green space. A very wide range of land uses are included in 'green space', including all agricultural land uses and all types of open space. Domestic gardens comprise a further 14.6 per cent of the land area of the city giving a total of 70.3 per cent of open space within the city (Table 10.1).[7]

Table 10.1 Land use in Salford

Land Use Categories, simplified	Area for each category in 2005 (ha)	Percentage
Domestic Buildings	524	5.4
Gardens	1421	14.6
Non-Domestic Buildings	360	3.7
Road	1007	10.4
Rail	68	0.7
Path	74	0.8
Green space	5400	55.7
Water	177	1.8
Other	669	6.9
Not classified	0	0.0
TOTAL	9700	100.0

Source: *Communities and Local Government Land Use Statistics for Salford January 2005, updated 27th Jamuary 2007*

At a more detailed level, data from the city of Salford provides an audit of green space at both a city and community committee area.[8] Importantly, within the context of this paper, golf courses comprise 8.0 per cent and sports pitches a further 8.1 per cent of the open space of the whole city. Overall, about 64 per cent of the city can be considered as urban. Sports fields and golf courses comprise 22 per cent of the open space, space which could make a vital contribution to protecting and enhancing biodiversity within the city.

Broadening the discussion to the wider region around Salford – Greater Manchester – we see similar problems, opportunities and challenges. As in Salford, there has been a major decline in manufacturing industry, which has led to the development of service industries. These service industries are often located outside urban centres, in suburban locations. Changing economic activity from the declining manufacturing industries and the expanding service sector, has decentralised employment and population,

7 Communities and Local Government, *Land Use Statistics for Salford January 2005*. Available internet: http://www.neighbourhood.statistics.gov.uk/dissemination/LeadTableView.do?a=3&b=276781&c=salford&d=13&e=8&g=354179&i=1001x1003x1004&m=0&r=1&s=1254210295367&enc=1&dsFamilyId=1201 Accessed 28th September 2009.
8 City of Salford, *Salford Greenspace Strategy*, Salford: Salford City Council 2006.

which have brought housing, transport and development problems. The *Association of Greater Manchester Authorities*, a body which represents the ten Greater Manchester city and metropolitan councils, highlights the promotion of development and regeneration throughout Greater Manchester as a main priority for the region, with the aim to strengthen economies and improve physical and social infrastructure. The major economic growth and developments have been in suburban and periurban areas '…where open land has been easier, cheaper and more attractive to develop.'[9] Greater Manchester's countryside, habitats, wildlife, green space and brown-field sites are at risk from encroachment by industry, housing and business park developments and landfill sites. Woodland planting schemes and recreational usage also threaten Greater Manchester's habitats and wildlife.[10] Here, as in Salford, the growth of golf courses in recent years raises questions over the impacts of open space used for sporting activities within the context of the landscape ecology of the urban area and more widely within the context of the sustainability of an area.

The growth of the golf industry

The number of golf course developments has dramatically increased since the mid 1980s (Figure 10.1).[11] As the number of golf course developments increase and, as more people turn to golf in their leisure time, the environment, economy and community role that golf courses play becomes significant. A report by the Golf Research Group based on a telephone survey of every golf course in the UK declared that there were 2,407 golf courses in the UK; 890 of these were classed as exclusive.[12]

The data in Figure 10.1. illustrates the number of courses, which have been built on greenfield sites. This number peaked in 1993. Since then, the number of new developments has declined. The average course had 529 members totalling approximately 1.17 million golf members in the United Kingdom in 2001, with a further 7.5 million people who really want to play golf. Although more people are interested in playing golf, unfortunately, many people in the UK are put off by the perception that golf is expensive (equipment and lessons), difficult to learn, takes too long to play, and is socially exclusive.[13]

To encourage more people to play golf throughout the UK, golf, following the United States model, is becoming relatively less expensive and more

---

9 AGMA, Association of Greater Manchester Authorities, *Review of Greater Manchester Strategic Planning Framework 1999*. Available internet: www.stockport.gov.uk/council/eed/planning/gmspf.pdf. Accessed on 14 September 2002.
10 *Countryside Agency Manchester Conurbation*. Available internet: http://www.countryside.gov.uk/LivingLandscapes/countryside_character/north_west/manchester_conurbation.asp. Accessed on 22 November, 2003.
11 D. Gilleece, *Growth Continues to Carry On, 1998*. Available internet: www.golf-research.com/irishpress.htm. Accessed on 7 August 2002.
12 Golf Research Group. *Financial Performance of UK Golf Courses, Golf Research. 2001*. Available internet: http://www.golf-research.com/index.htm. Accessed on 14 June 2002.
13 Golf Research Group. *Financial Performance of UK Golf Courses, 2001*.

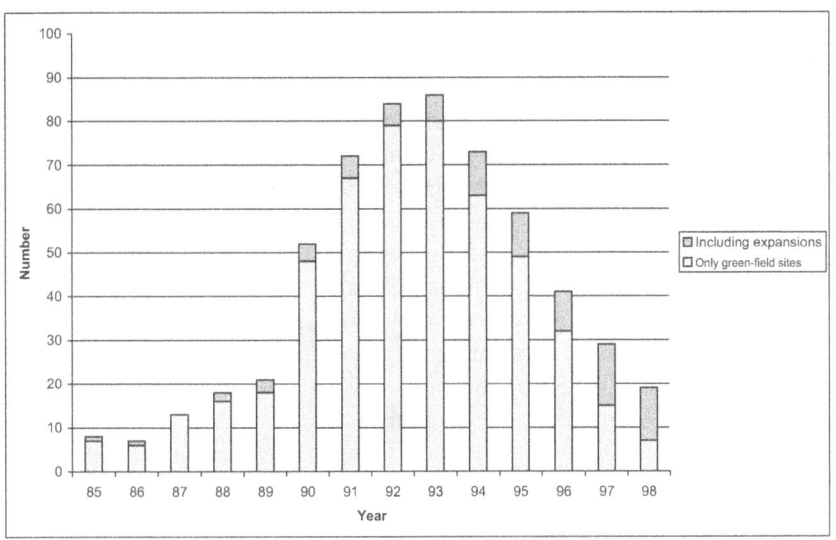

Figure 10.1 Number of new UK golf course developments built in 1985–1998
Source: Golf Research Group

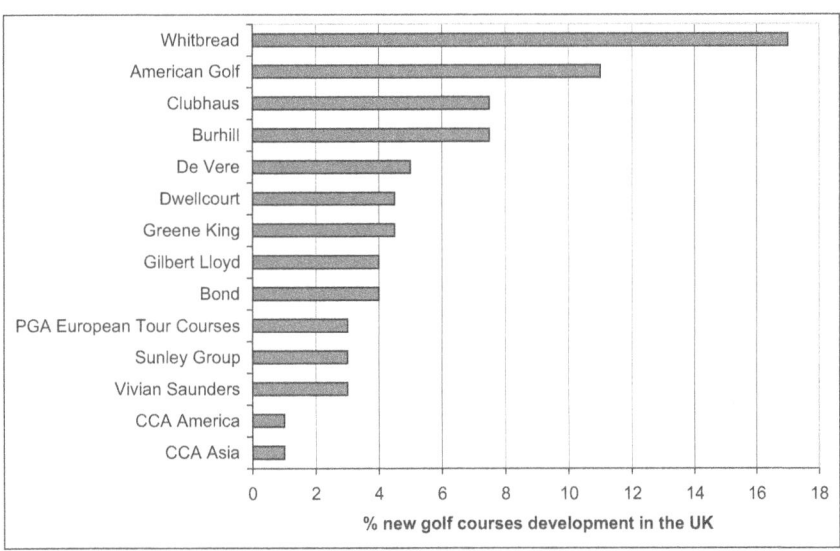

Figure 10.2 Percentage of new golf course developments, by multinational organisations, in the UK during the 1990s
Source: Golf Research Group. Financial Performance of UK Golf Courses, 2001.

egalitarian. In the 1990s there was an emergence of golf chains (see Figure 10.2). Golf companies owning multiple courses in the UK include Whitbread, American Golf Corp and Clubhaus. Golf course chains can offer a more affordable, accessible way into the game because of their deep pockets and economies of scale. It appears that commercial rather than social interests have driven these developments. Many of these courses, by incorporating health and fitness club facilities alongside the golf courses, are built to attract whole families, rather than just men. This style of development, which provides extra facilities for families, appears to be working. At the same time, the golf industry is subject to fluctuation in numbers and, hence, economic performance. There has been a strong increase in the number of developments since 1985 and, during the 1990s, a large proportion of these were completed by multinational organisations.

As golf courses occupy a significant area of cities they have the potential to be the lungs of urban, periurban, suburban living. If this is so, an increase in the number of golf courses would be beneficial to both the natural environment and the local community. To ensure environmental and community benefits, an environmental management programme for the golf course should be a priority. However, if the golf industry, like the tourism industry, is susceptible to economic fluctuations, during economic downturns the amount of money spent playing golf would be affected. This has implications for environmental management and environmental management plans, both of which involve money and a great deal of time, and under difficult circumstances implementation may discontinue. In other words, if golf courses are the lungs of urban living and their economic viability declines, then golf courses could experience the same fate as agricultural lands. At this point it may be useful to look at the development of one particular gold course, the Marriot Worsely Park, and examine the landscape ecology implications of its development.

Case-study: The Marriott Worsley Park

The Marriott Worsley Park (MWP) is a classic case of the reallocation of periurban agricultural land from agriculture to sport and recreational use. Worsley is recognised as one of the cradles of the Industrial Revolution. The presence of coal was recorded as early as the 1300s.[14] In 1759 an application was made to Parliament to construct a canal from Worsley to Manchester in order to transport cheap coal to the evolving powerhouse of Manchester.[15] This canal, the Bridgewater, became the first large-scale commercially viable canal in Britain. Coal mining ceased in 1887.[16]

---

14 N. Redman, *An Illustrated History of Worsley and the Marriott Manchester Hotel and Country Club,* Whitbread PLC 1998.
15 H. Wickham, *Worsley in the 18th century. A Study of a Lancashire Landscape*, Neil Richardson 1984.
16 A. Monaghan, *Duke of Bridgewater's Underground Canal at Worsley: Brief Introduction 1999.* Available at http://www.d.lane.btinternet.co.uk/canal3.htm Accessed 29 June, 2002

During the 1960s farmland around Worsely was turned into housing developments. Peel Holdings plc, who purchased Bridgewater Estates in 1984, initiated further housing programmes on farmland surrounding the village during the 1980s and 1990s. In 1987 Peel Holdings plc commissioned L&R Leisure Consultants to report on existing facilities in the area and to identify the most viable option for future development. Their final report proposed a hotel and leisure complex. This encouraged Peel Holdings plc to apply for planning permission. Plans were presented which incorporated the extant Worsley Old Hall, redevelopment of Old Hall Farm, and the construction of a golf course. In April 1990 the new plans were passed by Salford City Council.[17] The proposed regeneration of Worsley Old Hall and Old Hall Farm, which had been neglected since the tenants left in 1988 and had fallen into disrepair, encouraged residents to back Peel Holdings' plans. Finally, in 1995 Whitbread plc took over the site announcing a £15 million development and used their own landscape and architecture consultants to develop the site plans. Work began on the site in 1997; the hotel opened in 1998 and the golf course opened in spring 1999.

The MWP site can be conveniently separated into hotel and leisure centre buildings, the grounds surrounding the hotel and leisure centre, and the golf course. The grounds surrounding the hotel and leisure centre are largely manicured, ornamental landscaped gardens designed to minimise the impact of the hotel buildings and car parks on the landscape, and offer ornamental and seasonal interest.[18] The golf course and surrounding grounds are less ornamental, designed to provide a challenge and aesthetic interest for the golfers.

Initial development of the golf course involved heavy construction and intense land movement. Topsoil was removed (a condition of the planning application was that no soil was allowed to leave the site), sub-soil was shaped and the topsoil replaced to a depth of approximately 100–150mm to create an undulating landscape which is evident today. An irrigation system was designed to collect rainwater which would be stored in a reservoir for use in the summer months. Soil dug out to create the lakes and reservoir was also used to create the undulating features of the golf course. In addition to the soil dug out on site, materials were brought from outside the MWP site to aid golf course design and create the current golf conditions. These materials were sand, gravel, root zone (a combination of sand and peat) and pre-germinated seeds (mixed with water and paper). Such materials affect plant growth by facilitating good turf growth and increasing the kind of turf condition which golfers aspire to play on.

Plantations were incorporated into the golf course design to integrate the course with the existing woodland, compensate for any loss of hedgerow and woodland, delineate fairways and provide a challenging golf course. Approximately 46,000 whips of mixed species were planted. Since planting, the plantations have been thinned as appropriate.

---

17 P. Nears, Personal communication, Strategic Planning Officer, Peel Holdongs plc, Peel Dome, The Trafford Centre, 23 October, 2002.

18 M.J. Pennington, *Worsley Park Landscape Report*, Southport: Mike Pennington Landscape Architect 1995.

In addition to scattered mature Beech (*Fagus sylvatica*), Sycamore (*Acer pseudoplatanus*), Common Lime (*Tilia x europaea*), Oak (*Quercus robur*) and Holly (*Ilex aquifolium*) which were found across the golf course, the MWP site contained six blocks of woodland. The oldest woodland, Rands Clough, appears on a map dated 1764. Rhododendron (*Rhododendron ponticum*) was spread throughout the majority of the golf course and was found in all but two of the woods. During the construction of MWP areas of woodland were destroyed, but fifteen additional plantations (of approximately 30 whips and trees) were planted to compensate for any loss.

Hedgerows, an important wildlife habitat, were found across the site. Prior to construction, hedgerows had not been managed since the last farming tenants left in 1988 and all were overgrown, with poor ground flora, and gaps throughout. Extant hedgerows were dominated by Hawthorn (*Crataegus monogyna*). There was no evidence of ancient hedgerows at the MWP site.[19] During golf course construction, one hedgerow was completely removed and five hedgerows were fragmented by footpaths, golf buggy paths and fairways.

To the south of the golf course a stream was constructed to increase the challenge of the golf course and for aesthetic purposes. Three remnant ponds were incorporated into the course design. These appeared eutrophic and overgrown with Greater Reedmace (*Typha latifolia*). During the growing season of 2001 a 10-meter buffer zone was established around the ponds to prevent further nutrient enrichment. Ponds are regionally important habitats.[20] All three ponds were reported to contain fish, Common Frog (*Rana temporaria*) and Common Toad (*Bufo bufo*). Smooth Newt (*Lissotriton vulgaris*) and Palmate Newt (*L.helveticus*) have been sighted on the golf course. In addition, there are four wetlands, three permanent and one seasonal, on the MWP site. Common species in these wet areas include Greater Reedmace (*Typha latifolia*), Yellow Flag Iris (*Iris pseudocorus*), Common Duckweed (*Lemna minor*) and Water Thyme (*Elodea* sp.).

Since construction, and subsequent play and management, the MWP greenkeepers have incorporated an environmental management regime which encourages sward growth and reduces chemical applications throughout areas of rough, limits the amount of alien species such as Rhododendron (*Rhododendron ponticum*), encourages buffer zones around waterbodies and supports natural regeneration within woodland and the woodland edge.

The MWP is bordered by residential housing to the east, south and west, and a transport corridor to the north. Walkden Road, to the east, is predominantly detached housing with large gardens and large mature trees; additionally there are a number of terraced and semi-detached houses with smaller mature gardens. To the south, Leigh Road, has a mixture of larger and smaller detached properties all with mature gardens. Similarly to the west Randsclough and Stechworth Streets contain semi-detached housing

---

19 P. Tattersfield, *Worsley Park: An Appraisal of its Nature Conservation Value and Implications for Development,* Manchester: Penny Anderson Associates 1990.
20 *The Urban Mersey Basin Natural Area: A Nature Conservation Profile,* Peterborough: English Nature 1997.

with mature gardens. The importance of urban gardens has been long recognised. However, due to access difficulties they remained an understudied aspect of our natural heritage. A project lead by the University of Sheffield – Biodiversity in Urban Gardens – has gone a long way to address this issue. An important finding from that project is that domestic gardens can form extensive, inter-connected tracts of green space. Therefore, as for other categories of urban green space, the benefits of individual gardens probably arise from their role as isolated patches, as components of a landscape that includes other vegetation or as corridors through the urban matrix. The most important factor determining the biodiversity value of a garden is the structure of the vegetation.[21] Hence as many of the gardens surrounding the MWP are described as mature (with the majority classed as medium-large gardens) it could be assumed that these gardens may be biodiversity rich and/or offer value as corridors and feeding grounds to species or habitats associated with the MWP and with the wider Salford community area as a whole.

Findings from the case-study

The analysis of land use within the city of Salford presented in this chapter clearly illustrates the importance of sports pitches and golf course in terms of the area of the city devoted to these uses. Open space strategies for the city demonstrate that they concentrate on the public open space and seem to overlook the large areas which are in private ownership. This is an important omission because whilst planners are not able to dictate practices in existing developments they are able to make recommendations in the context of new developments which could significantly affect the resultant structure and functionality of a development.

Analysis of the habitats created and the species of plants incorporated into the design of MWP indicates that many of the features of the wider landscape are incorporated into the golf course. It is important to recognise that within a golf course much of the area is not used for golf: as little as 10 per cent of the area needs to be intensively managed leaving 90 per cent available as wildlife habitat.[22] Hence, although difficult to quantify precisely, it is clear that the nature resource of a city is much larger than those publicly owned sites which are addressed in green space strategies and extends into land in private ownership of which gardens and golf courses are important components.

This leads to a new model for the creation of urban ecological frameworks. The model emphasises that nature conservation gains are achieved by recommending general measures (using the most up-to-date studies of best practice available) that can be used to enhance biodiversity wherever and whenever possible. Species populations will benefit from this approach but

---

21 R.M. Smith, K.J. Gaston, P.H. Warren and K. Thompson, 'Urban Domestic Gardens (V): Relationships Between Landcover Composition, Housing and Landscape', *Landscape Ecology* 20 (2005), 235–253.
22 B. Tobin and B. Taylor, 'Golf and Wildlife', *British Wildlife* 1996, 7 (3) 137–146.

in a much less structured and guided way than by formally designating or creating a framework comprising core areas, nature restoration zones, buffer zones and corridors. By adopting the policy proposed here, recognising a great diversity of public and private green spaces, a sustainable framework emerges from the 'ground up' rather than from the 'top down'.

One outcome of this model is a 'toolkit' recommending the best ways of improving urban areas for biodiversity in general terms, together with some spatial information about where the tools in the kit should be put to best use. In a sense this framework model also seeks to establish an informed 'frame of mind', encouraging measures to be taken to enhance biodiversity wherever possible.

Conclusion

As post-industrial cities continue to develop it is likely that greater emphasis will be placed on the open spaces of these cities. City residents and those who work in cities expect attractive environments in which to live and work. Open spaces and contact with nature have an important role to play in the quality of life of all users of cities. Sports fields are important green spaces which, in the UK, are often privately owned and are developed for uses other than wildlife. The study presented here has quantified that resource for one city in the northwest of England, explained the historic developments which have lead to this resource and demonstrated that it is possible to develop a framework into which these sites can be incorporated but within which they maintain their important, distinctive roles.

CHRISTIANE EISENBERG AND REET TAMME

11 The Golf Boom in Germany 1980–2006: Commercialisation, Nature Protection and Social Exclusion

Since the end of the twentieth century many countries around the world have experienced a golf boom. Seen from the point of view of demand, this involves the adoption of the general standards of the game communicated by the media and worldwide tourism, not only regarding the rules, but also with respect to clothing, equipment and 'ambience'. Seen from the supply side, we can observe a rapid growth in golf courses.[1] By contrast with earlier phases of development in golf, these are no longer primarily built by select and fashionable clubs, but rather by hotel chains and real estate investors who build luxurious American-style condominiums to accompany them. Within these projects, golf courses then become a sort of domestic lawn. This has a positive side effect because the very existence of golf courses prevents any further building activities and thereby raises the value of the real estate. However, since golf course architects are duty-bound to comply with international golfing standards, they often come up with landscape concepts which are completely alien to the local topography and distort the indigenous flora and fauna.[2]

The international golf industry has been made sensitive to ecological issues, not only by the global operations of the anti-golf movement but by other factors, and has now included ecological viewpoints in its activities, which also comprise the construction of golf courses (course design, grass seeds and so on).[3] In addition, the observance of ecological standards when

---

1 B. Stoddart, 'Golf international: considerations of sport in the global marketplace', in R.C. Wilcox (ed.), *Sport in the Global Village*. Morgantown: WV Fitness, 1994, 21–38.
2 J. Bale, *Landscapes of Modern Sport*. Leicester: Leicester University Press 1994, 21–38. See also J. Pern, 'Golf courses and the built environment', in European Institute of Golf Architects (ed.), *EIGCA Yearbook 2006*, 43–45
3 D. Stubbs, 'Environmental issues facing golf in Europe', in *European Institute of Golf Course Architects*, http://www.eigca.org/articles1.php (accessed 1 August 2006). See also G. Fierz, 'Golf und Tourismus. Zum Vierten Internationalen Anti-Golf-Tag vom 29 April 1996', and A. Pleumarom, 'Golf und Tourismus in Südostasien: Kampf um Land- und Menschenrechte. Zum Vierten Internationalen Anti-Golf-Tag vom 29. April 1996' (both articles are available in the newspaper clipping collection of the Deutsches Golf-Archiv an der Deutschen Sporthochschule Köln, Nr. 2108), as well as M. Rodrigues, 'The political sociology of golf in South Asia', in http://mail.sarai.net/pipermail/reader-list/2005-May/005728.html (accessed 3 Dec. 2006).

building golf courses has become an issue *sui generis* in Europe. Whether, how far, and with what effects this issue is acknowledged is, however, dependent on the particular local situation.[4] In the case of Germany we must take into account that, when the golf boom set in at the end of the 1980s, it was in a context where certain kinds of mis-use of open space were no longer possible (see Chapter Seven by Jürgen Breuste. In the late twentieth century, nature protection movements and environmental organisations, widespread public opinion and officials had developed a high level of sensitivity to the possibility of damage and destruction to the natural environment, so that certain standards became anchored in laws and regulations which had to be adhered to by golf investors, amongst others.

In this chapter we would like to outline how the golf boom at the end of the twentieth century fitted into this context, and was 'tamed' by it. In doing so we shall first outline the quantitative development and certain structural characteristics of golf in Germany in the second half of the twentieth century. We shall follow this with an analysis of the situation in the capital, Berlin, and the surrounding federal state of Brandenburg where the effects of the boom were exceptionally intense, because the collapse of the German Democratic Republic (GDR) and German reunification in 1989/90 amplified general global trends. We shall conclude by outlining the complex of relationships which has grown up between golfing and nature protection interests, and discuss the special constellation of golf today, which might also develop into a political issue in contemporary German society.

Constraints on the development of golf in Germany

Golf arrived in Germany from the British Isles as early as the end of the nineteenth century. That said, until around the 1980s, it remained the exclusive domain of a small elite of industrialists, bankers and other very affluent and rich persons and their families. Seen from a sport history perspective, golf was and is still a fringe phenomenon.[5] All the more so because – unlike genuinely popular sporting disciplines in Germany such as football, athletics, shooting, swimming and cycling[6] – golf received no public funding between the two world wars (with the exception of a short period in the Third Reich). In the German Democratic Republic (GDR), which was founded in 1949, the sport was heavily repressed right from the very start; not only on ideological grounds, but also because the 14 surviving golf courses were earmarked for military exercise areas and industrial plants. When the GDR collapsed in 1989, golf in East Germany was to all extent

---

4 In this respect the golf boom, like other global developments, is a phenomenon of glocalisation.
5 C. Eisenberg, *"English sports" und deutsche Bürger. Eine Gesellschaftsgeschichte 1800–1939*. Paderborn: Schöningh 1999, 156, 209, 330; Deutscher Golf Verband (ed.), *100 Jahre Golf in Deutschland*, 4 vols. Oberhaching: Albrecht Golf Verlag 2007.
6 C. Eisenberg, 'Massensport in der Weimarer Republik: Ein statistischer Überblick', *Archiv für Sozialgeschichte* 33 (1993), 137–178.

and purposes dead, not least because it had been impossible to purchase equipment, golf clubs and balls.[7]

Neither was the sport more than a minor phenomenon in the Federal Republic of Germany (FRG) before 1989. To a certain extent this was due to the fact that West Germans had developed completely different sporting preferences to which they held fast. Golf was dismissed as being boring, and was widely regarded as an old people's sport. But the main reason for its lack of popularity was that, even in an affluent society, only a very small portion of the population wanted to, or could afford to, join a golf club because there were so many other opportunities for spending money: housing, motorcars, holidays and the like. Since golf was regarded as a sport for rich people, it continued to receive no public funding – except for certain tax concessions accorded to non-profit making clubs.[8] This denial of funding was one of the main reasons why the (West) *German Golf Association* (DGV) decided to join the *Deutscher Sportbund* (DSB, German Sporting League), even though it played a completely minor role compared to football, athletics, swimming, gymnastics, and a huge number of other sports. Another reason why there was no wider public promotion of golf clubs in the Federal Republic (as there was, for example, in the Netherlands, Sweden and Great Britain)[9] was that most clubs only allowed their own members and guests to use their courses. Municipal golf courses did not exist in the old Federal Republic, with the single exception of a course in Düsseldorf (a concession to the golf-crazy Japanese community in the city); in 1984 however, even this course was closed down as a result of a budgetary crisis.[10] By 1987, after a period of slow but steady growth, West Germany had a total of 238 clubs, most of which possessed 18-hole courses and a minority had nine or six-hole courses; but almost 30 clubs did not even have their own golf course. Average membership figures were under 600 persons per course; according to experts this was regarded as the absolute minimum in order to be able to finance an 18-hole course.[11]

This gradual growth speeded up towards the end of the 1980s. From then until 1998 there were around 30 new golf courses created every year. Even after the late 1990s, when a certain degree of saturation had already been reached, there were still around 20 new courses a year.[12] By 2005 the total number of golf courses registered with the *DGV* had reached 648. In addition there were a small number of non-registered golf clubs and courses. We have only been able to find exact figures for 2001, when there were 34 such courses. Most of them had six to nine holes: eight were pure driving ranges.[13]

---

7 R. Baartz, *Der Konflikt zwischen Sport und Umwelt, dargestellt am Beispiel der Entwicklung des Golfsports im Raum Brandenburg-Berlin*. Stuttgart: Franz Steiner Verlag 1994, 108.
8 For more precise information see A. Steinbrück, *Strukturen des Golfmarktes in Deutschland: Potentiale, Einflussfaktoren und internationaler Vergleich*, PhD thesis, Deutsche Sporthochschule, Köln 2004, 192 ff.
9 P. Laurer, *Die Ausbreitung von Golfanlagen in Bayern und Ostdeutschland. Innovation – Diffusion – Erfolgsfaktoren*, PhD thesis, Erlangen-Nürnberg 2001, 126.
10 Steinbrück, *Strukturen*, 69.
11 Baartz, *Konflikt*, 104–105.
12 Steinbrück, *Strukturen*, 142 f.
13 Steinbrück, *Strukturen*, 143, 235.

Figure 11.1 Public Golf Eastside Berlin. *(Photograph: Reet Tamme).*

Thus the total number of golf courses in Germany rose from 238 in 1987 to around 700 in 2005, an increase of about 300 per cent[14] and the number of registered golfers, which had been around 96,000 in 1987, rose by about 500 per cent. In 2005 this number had exceeded the 500,000 mark.[15]

As a result of the rise in the number of courses which can be largely attributed to the investment activities of commercial companies,[16] Germany advanced strongly in international golf statistics. Measured in terms of the proportion of golfers in the country to the population as a whole, in 1992 Germany was in 11th position in Europe;[17] but by 2001 it had risen to third

---

14 Similar conclusions are arrived by Steinbrück, *Strukturen*, 143.
15 'DGV-Mitglieder und Golfer in Deutschland 1907–2005', in Deutscher Golf-Verband (ed.), Statistiken: Nationale und regionale Golfsport-Entwicklung. http://www.golfparadise.com/cms/aktuelles/2006_02_02_DGV_Statistiken.php (accessed 28. Nov. 2006).
16 Today at least 60 per cent of golf courses in German are the property of such societies or are under their management; Steinbrück, *Strukturen*, 75; F. Billion, 'Golf resorts: Freizeiträume für privilegierte Minderheiten oder Beitrag zur Attraktivitätssteigerung deutscher Tourismusregionen?', in P. Reuber and P. Schnell (eds), *Postmoderne Freizeitstile und Freizeiträume. Neue Angebote im Tourismus*. Berlin: Erich Schmidt Verlag 2005, 329–337, here 331.
17 If England, Scotland and Wales are counted separately, even 13th place; Baartz, *Konflikt*, 115.

place behind Great Britain and Sweden. If we only compare golf courses and practice ranges, Germany was actually in second place, in front of Sweden. That said, Great Britain which had around 2,800 golf courses in 2001 (almost 4 times as many as in Germany) still has a huge lead.[18]

The situation in Berlin and Brandenburg

Seen as a whole, the golf boom in Germany is a phenomenon which has been restricted to the 'old' federal states (former FRG), which still contain over 90 per cent of the golf courses today.[19] The majority of courses are concentrated in urban centres like Hamburg, Munich, Frankfurt and the Ruhrgebiet. Furthermore, in Bavaria, which has a strong tourist industry golf courses can also be found in and around smaller towns, and in the countryside. The fact that the 'new' federal states (former GDR) continue to lag behind requires an explanation, in so far as it might be expected that there would have been a much larger potential for golf in the *tabula rasa* situation which followed reunification, and because conditions for investment were also remarkably attractive for three main reasons. Firstly, the price of real estate was extraordinarily low because the collapse of agricultural activity in the GDR had led to a considerable decrease in the amount of land being used for such purposes. Land prices in Mecklenburg-West Pommerania and Brandenburg, for example, were five to eight times lower than in West Germany, especially in Bavaria, the golf region par excellence.[20] Secondly, towns and local authorities, whose industries had collapsed because they were no longer competitive, had a huge interest in co-operating with investors of any sort. Thirdly, a new law, the so-called *Fördergebietsgesetz*, passed in 1990 for the whole of the former GDR region, provided for tax privileges for investors, including those planning to build golf courses. This also made golf courses interesting investments as depreciation projects.[21]

The prospects for profitable golf investments in Berlin seemed even rosier. In the first place, thanks to the Wall, the topography of West Berlin was rather like an island, and the lack of any markets for land in the GDR had meant that the old pre-war residential structures outside the city boundaries had not changed and there had been no process of suburbanisation. As a result, after reunification, there were broad areas of countryside immediately adjacent to built-up areas. Around 1989 the proportion of residents in the Berlin city centre compared to those in the adjacent countryside was 80:20, whereas in cities like Hamburg and Munich it was 50:50.[22] Furthermore,

---

18 World Amateur Golf Council, Application to the IOC, April 12, 2001, Appendix A, here op cit. Steinbrück, *Strukturen*, 162.
19 According to figures in: 'Golfplätze nach Landesgolfverbänden 2005', in Deutscher Golf-Verband (ed.), Statistiken.
20 Laurer, *Ausbreitung*, 196.
21 Laurer, *Ausbreitung*, 193–194.
22 W. Beyer and M. Schulz, 'Berlin – Suburbanisierung auf Sparflamme!?', in K. Brake and J.S. Dangschat and G. Herfert (eds), *Suburbanisierung in Deutschland: Aktuelle Tendenzen.* Opladen: Leske & Budrich 2001, 123–150, here 123.

the city of Berlin lagged behind in its number of leisure facilities outside built-up areas. It was a well known fact amongst golfers that the only two clubs in West Berlin before the fall of the Wall had long waiting lists, and for this reason many West Berliners were members of golf clubs in other parts of West Germany.[23] No less significant, the decision to move the government from Bonn to Berlin made the prospects for golf investors even better, because a potential golf clientele would be moving into the city.

It is no surprise that the authorities in Berlin and the adjacent federal state of Brandenburg received 73 applications to build golf courses in the years up to 1998, and between 1990 and 1992 eight new golf courses were opened.[24] Looked at in the longer term however, this development was not as successful as it seems. 38 of the 73 applications were withdrawn, and most of the other projects came to nothing.[25]

There were a variety of reasons for this. Firstly, the high hopes in an upturn in the economy in East Germany collapsed within a short period of time, with the result that there was an exodus of residents from the 'new' federal states. Berlin too failed to attract new inhabitants; indeed its growth was below average. Since there were plenty of rooms to let at moderate prices in the city centre, the suburban sprawl did not occur as quickly as originally expected.[26] As a result, there was no major increase in the demand for golf. Secondly, the thin line between profit making and eligibility for tax relief for non-profit making golf club projects (the usual pattern of golfing organisation in Germany) was difficult and risky.[27] Even the opportunity introduced by the government in 1995 for club members to invest personally in golf projects and thereby benefit from tax relief, failed to change the situation fundamentally.[28] The only real new ray of hope for golf investors proved to be a programme of urban renewal in the east (the so-called *Stadtumbau-Ost-Programm*) which provided funds for rebuilding industry and housing in the 'old' federal states. However, only driving ranges and smaller urban areas could benefit from this.[29]

The third and principal reason for the abrupt end to the wild and unregulated boom in golf courses in the first half of the 1990s was the introduction of lengthy application procedures (which might last from one to three years),

23 Laurer, *Ausbreitung*, 211.
24 Laurer, *Ausbreitung*, 200, 214.
25 Laurer, *Ausbreitung*, 214.
26 Beyer and Schulz, 'Berlin', 123.
27 G. Bentz and K. Gieseler and K. Cachay, 'Lässt sich Sport verkaufen? Probleme der Kommerzialisierung', in K. Gieseler i.A. des Deutschen Sportbundes (eds), *Menschen im Sport 2000*. Schorndorf: Hofmann 1988, 215–233.
28 For a detailed treatment of this problem see Steinbrück, *Strukturen*, 192–202.
29 Good-Practice-Beispiel. Koordinierungsstelle Flächenmanagement Berlin, Marzahn-Hellersdorf (ed.), in: Stadtumbau Ost – Stand und Perspektiven, Erster Statusbericht der Bundestransferstelle i.A. des Bundesministeriums für Verkehr, Bau und Stadtentwicklung und des Bundesamtes für Bauwesen und Raumordnung, 55, in: http://www.irs-et.de/download/Erster_Statusbericht_Stadtumbau_Ost.pdf (accessed 15 May 2006); Chronik der Freiraumentwicklung, in: Senatsverwaltung für Stadtentwicklung, http://141.15.4.17/umwelt/landschaftsplanung/chronik/#2000 (accessed 26 Nov. 2006).

and the costly stipulations imposed by the city authorities.[30] Whereas in Brandenburg, for example, there were no major conservation laws in existence immediately after reunification, and local authorities initially tended to welcome every offer of investment with open arms, the authorities soon began to scrutinise projects for their effects on the environment.[31] In addition, they increasingly began to apply West German standards to their activities with regard to nature protection and monument protection, and these were too high for many golf course investors. Conflicts arose between investors and the authorities particularly with regard to questions of ecological, natural and monument protection, because investors in general did not wish to build on redundant agricultural areas, and preferred to concentrate their activities directly on protected landscapes, or areas in their immediate vicinity.[32]

The effectiveness of official regulations becomes very clear, when we look at the specific areas which were used for building new golf courses in Berlin and the surrounding countryside in the years immediately following reunification; and at the two time spans 1990–92 and 1993–96 separately. Of the eight courses which were opened between 1990 and 1992,[33] one was built on the site of a former agricultural co-operative (in East Germany, an agricultural co-operative was a large concern with intensive and presumably monostructural use of land); two were built on large orchards; two more were built in 'locations of natural beauty' and in local recreational areas (we suspect that these were probably landscape and nature protection areas); and there is no information on the remaining three courses.

Four further courses were built between 1993 and 1996. By contrast these appear to have caused no major problems with regard to nature and environment protection. One course (now closed) was on a disused aerodrome; [34] one 36-hole course for tourists was built on an old agricultural and forestry area (the *Sporting Club Bad Saarow,* 250 ha); a similar course was made on a former agricultural area (the *Seddiner See Country Club* in

---

30 Steinbrück, *Strukturen*, 166. Applications to build golf courses in Germany have to go through four stages: regional planning, building plans, building authorisation and expert authorisation. The first two stages comprise an urban and regional planning process and an application to change area usage plans (here citizens and the local authority are involved) and the drawing up of a building plan. The other two stages comprise the completion of a detailed building application, which also has to contain a plan for maintenance and development, as well as applications for re-forestry, land clearance, exemption from landscape protection regulations, and finally applications for water rights for building springs, ponds etc. Cf. K.F. Grohs and R. Preißman, 'Genehmigung von Golfanlagen', in Pfaff Marketing (ed.), *Golf Management Handbuch*. Sinn-Edingen: Pfaff Marketing GmbH 1996, 3–17; Laurer, *Ausbreitung*, 102.
31 Baartz, *Konflikt*, 23, 54, 57 f.; Laurer, *Ausbreitung*, 192; B. Meinhard, 'Die Enklaven', *Süddeutsche Zeitung* 21.10.1992 (quoted from the newspaper clipping in: Deutsches Golf-Archiv (Deutsche Sporthochschule Köln, Reg. Nr. 441).
32 Baartz, *Konflikt*, 45, 47, 146 and passim; C. Berg, D. Egge and G. Hartmann, *Golfplätze in brandenburgischen Raumordnungsverfahren: Beitrag zur Bedarfsermittlung und verwaltungsjuristischer Fachbeitrag. Gutachten im Auftrag des Landes Brandenburg*. Hannover: Büro für Tourismus und Entwicklungsplanung, BTE 1992, 35–36.
33 Laurer, *Ausbreitung*, passim, and own research.
34 Laurer, *Ausbreitung*, 202.

Figure 11.2 Berliner Golfclub Stolper Heide. *(Photograph: Reet Tamme)*

Wildenbruch, 230 ha);[35] and one course was sited around a long section of the old Berlin Wall on a landscape protection area (*GC Stolper Heide*). It is worthy of note that both 36 hole courses were built by particularly wealthy investors: the Kempinski hotel chain (which has since sold its course to another company), and a group of investors around Prince Ferdinand von Bismarck.[36] The course attached to the Stolper Heide golf club was also strongly connected with real estate activities.

All in all there are now 17 golf clubs in Berlin and Brandenburg, each of which possess at least one 18-hole course. Three of these clubs lie within the city border. In addition the city has five to six smaller golf courses (some of them are only driving ranges), which are open to all players for a fee.

---

35 Laurer, *Ausbreitung*, 202. Clearly this course had had different previous uses. For a time it also seems to have been a landscape protection area: cf. J. Götting-Frosinski, 'Wohnen am Golf', *Grünstift* H. 3 (1994), 8–11.
36 Götting-Frosinski, 'Wohnen', 8. For further projects of this type in Germany see F. Billion, 'Golf Resorts'.

Golf and nature protection

The restrictive policies of nature protection and planning authorities hindered a large number of golf projects in the 'new' federal states. But the conditions in the other federal states were no different. Here too golf projects were only authorised when they conformed to certain standards.[37] What sort of standards were they? And how did they come into being? In order to answer these questions, it is necessary recall that in 1976 the Federal German government passed a new Nature Protection Law (*Bundesnaturschutzgesetz*). Whereas the law till then had aimed at protecting individual flora and fauna, animals and 'nature and monuments', the new law was much more wide ranging (see also Chapter Seven by Jürgen Breuste). It demanded that 'nature and landscapes... in housing and non-housing areas are to be protected, maintained and improved in order to ensure that 1) the reproductive capacities of the natural world, 2) the usability of natural goods, 3) flora and fauna and the animal world, as well as 4) the variety, uniqueness and beauty of nature and the landscape are secured on a long-term basis as the living foundations for people and the prerequisite for their recreation in nature and landscape'.[38]

In order to fully understand the problem, it is important to keep in mind that the new Nature Protection Law simply gave the national government a framework of competence, whilst giving the federal states the responsibility of setting up their own individual plans for implementing the law. Furthermore, it explicitly emphasised people's recreational needs. Both these points together meant that conservationists had much more powers of intervention, a fact which placed them on a collision course with the land usage interests of industry and agriculture. Amongst other measures, the Nature Protection Law of 1976 defined national parks: on the one hand, they were supposed to serve the recreational needs of the people, but, on the other hand, they were also intended to help protect biodiversity, that is to lay down larger areas of biotopes and nature protection within buffer zones, thereby networking them to make them accessible to the general public, without disturbing or destroying their natural development.[39] From the 1980s individual federal states adopted the idea of 'biotope complexes' into their development plans.[40]

Against this background, it is no surprise that sporting bodies interested in using free open spaces began to take an interest in the concept of biotope complexes.[41] In this context, at the start of the 1980s, golf representatives

---

37 Laurer, *Ausbreitung*, 102–112.
38 Quoted from K. Ditt, 'Naturschutz und Tourismus in England und in der Bundesrepublik Deutschland 1949–1980: Gesetzgebung, Organisation, Probleme', *Archiv für Sozialgeschichte* 43 (2003), 241–274, here 265. Cf. also K. Ditt, 'Anfänge der Umweltpolitik in der Bundesrepublik während der 1960er und 1970er Jahre', in M. Frese et al. (eds), *Die 1960er Jahre als Wendezeit der Bundesrepublik: Demokratisierung und gesellschaftlicher Aufbruch*. Paderborn: Schöningh 2003, 305–347.
39 Ditt, 'Naturschutz', 266; H. Plachter, *Naturschutz*. Stuttgart u. Jena: Gustav Fischer Verlag 1991, 343–349.
40 Ditt, 'Naturschutz', 270.
41 Cf. the chapter entitled 'Sport und Umweltschutz' in W. Engelhardt, *Beharrlich in kleinen Schritten: 50 Jahre Natur- und Umweltschutz in Deutschland*. Berlin: Erich Schmidt Verlag 2002, 86–92.

made contact with a landscape ecologist named Professor Wolfgang Haber, who was promoting the concept of 'landscaped golf courses' at this time.[42] Haber had conducted a survey of traditional golf courses in the English county of Kent, and noticed that British nature protection authorities had elevated some of them to the rank of 'Sites of Special Scientific Interest' (SSSI) because of their huge variety of flora and fauna, some of which had grown up there over a time span of more than 100 years. Since there was no meshing of golf and nature protection interests in Germany at the time, he outlined a concept for designing golf courses, which attempted to create a coexistence between the greens (that needed a lot of care and attention) and the roughs, the patches of bushes and trees which lay between them. On the one hand, the roughs were intended to present challenges to the players; on the other hand they would be deliberately used to create biotopes.[43] Although parts of this concept were sharply criticised for their vagueness, the majority of golf course architects, golf course feasibility experts and other specialists in Germany took it up in the following years.[44]

The same was true of the *German Golf Association*, the *DGV*, which released its first leaflet on 'Golf Biotope Networking' to its own members and members of the general public in 1993, and put forward another very robust joint document with the Federal Office for Nature Protection in 2005. This document described golf courses as 'buffer zones between protected areas and the adjacent landscape' and advertised golf as the quintessential agent of nature protection. In this way, for example, water obstacles and ponds would operate – over and above their sporting functions – as valuable sanctuaries for the local animal world if they were designed to meet the needs of biotopes; and roughs could 'become meadows with a large variety of wild flowers and plants, given the right choice of seed and care'. 'Habitats for reptiles and small mammals' could also be created. The overall idea was to recognise and exploit the 'biotope potential' of golf courses.[45]

What was the purpose of this initiative? What was – and still is – its political function? Of course, first of all the initiative was aimed as a broadside against the growing number of golf course owners and golfers in Germany who favoured artificial landscapes along American lines, and the concomitant

---

42 W. Haber, 'Argumente für einen aktiven Naturschutz. Zur landschaftsökologischen Beurteilung von Golfplätzen', Special issue of *GOLF magazin*, März 1983; Deutscher Golf Verband e.V. (ed.), *Golf: Sport in Landschaft und Umwelt*. Wiesbaden: Albrecht Golf Verlag 1985, 4–9.

43 W. Haber, 'Argumente'; W.Haber, 'Golfplätze aus der Sicht des Naturschutzes', *Jahrbuch für Naturschutz und Landschaftspflege* 38 (1986), 129–135. For the SSSIs on British golf courses cf. I. Dair and J.M. Schofield, 'Nature conservation and the management of golf courses in Great Britain', in A.J. Cochran (ed.), *Science and Golf. Proceedings of the First World Scientific Congress on Golf, University of St Andrews, St Andrews, Scotland, 9–13 July 1990*. London: Spon 1990, 330–335; R.J. Price, 'The environmental impact of golf-related developments in Scotland', in Cochran (ed.), *Science and Golf*, 321–329.

44 E. Bierhals, 'Einordnung des Golfplatzes in Natur und Landschaft', in *Golfplätze im Trend*, (Schriftenreihe: Sport- und Freizeitanlagen J3/88). Bonn: Bundesinstitut für Sportwissenschaft 1988, 53–67.

45 Bundesamt für Naturschutz (BfN) and Deutscher Golf-Verband (DGV) (eds), *Biotopmanagement auf Golfanlagen*. Bonn: Bundesamt für Naturschutz 2005.

'global tendency towards aesthetic standardisation'.[46] But there was another aim – and we suspect this was the real aim of the campaign for 'golf biotope networking': the *German Golf Association* wished to promote itself as a competent adviser for local authorities and a potential recipient of funding, with respect to the clear 'international trend towards public golf courses'[47] in Sweden, Canada and the USA. For it can in no way be taken for granted that commercial golf course owners, who currently dominate the business of building golf courses and who will control the business in the future, will automatically co-operate with the *DGV*.[48]

Since it is expected that the concept of 'golf biotope networking' jointly developed with the Federal Office for Nature Protection will affect authorisation standards in the future, the *DGV* in its position as the lobbyist for the clubs, has it in its power to raise the official hurdles for purely commercial golf investors. In this way it can simultaneously prevent potential and current members of affiliated clubs from taking up their 'exit' options and joining forces with their commercial competitors. For, as with other German sports clubs and bodies, there is a growing lack of enthusiasm for joining golf clubs.[49] Since the implementation of 'golf biotope networking' requires the employment of caretakers on a permanent basis, this will also raise market hurdles for commercial competitors. Whereas golf club members can take on such duties in an honorary fashion, owners of commercial courses will be forced to engage and pay for extra staff.

Finally, by operating in accordance with the 'golf biotope networking' concept, golf clubs affiliated to the *DGV* are, to some extent, insuring themselves against financial collapse. Whereas commercial golf course owners with only a limited amount of capital backing would disappear from the market, members of clubs who went bankrupt might be able to continue to care for the course through another form of organisation. There would even be a possibility that local authorities would officially transfer future responsibilities to them. For it would be difficult to imagine a feasible alternative to organised golf for open spaces which had been turned into biotopes in conformity with the German Nature Protection Law. In this case, however, it would be impossible to justify restricting the use of such areas to members only.

---

46 *Biotopmanagement*, 21 f., quotations 21.
47 *Biotopmanagement*, 22.
48 As far as the DGV is concerned, a change in their statutes made in 2003 means that there is no longer anything to stop commercial golf course owners from becoming members of the association.
49 As in many other sports, there is a trend towards more individualism and unorganised activities in golf while the numbers of club members are becoming smaller; cf. *Golfmarkt der Zukunft 2005: Eine empirische Studie der GTC – Golf & Tourism Consulting in Kooperation mit dem Deutschen Golf Verband e.V.*, München: Süddeutsche Zeitung 2005, 166; Steinbrück, *Strukturen*, 146. For the general trend in German sports clubs see S. Braun, 'Leistungserstellung in freiwilligen Vereinigungen. Über 'Gemeinschaftsarbeit' und 'die Krise des Ehrenamts', in J. Baur and S. Braun, *Integrationsleistungen von Sportvereinen als Freiwilligenorganisationen*. Aachen: Meyer & Meyer 2003, 191–241.

The social and political consequences of the ecological golf business

It is uncertain whether the general public will be the beneficiary of the ecological taming of the golf boom in Germany. For golf courses that include biotopes cannot logically be used as mass venues. Looked at from a social and political perspective, we can therefore conclude that biotope complexes on golf courses are likely to encourage a growth in the privatisation of recreational areas. Up until now the general public in Germany, by contrast with several other countries, has been guaranteed free access to recreational areas. Taken to its logical conclusion, the developments in golf means that this principle will be abandoned – a negative side-effect which has not yet been a subject of serious public discussion in Germany. Whether this is due to the fact that ecological debates help to prevent this theme from coming out into the open, or whether the general trend towards the commercialisation of citizens rights via fees has had a 'habituation' effect, is neither here nor there.

Looked at from the perspective of sport history the development in golf, as outlined here, has also thrown up a few of its own problems. For it is difficult to reconcile the idea of sporting competition with the concept of 'golf biotope networking'. This is not only because competitions which draw large audiences inevitably produce a higher level of wear and tear (if only temporary) on the course. There is an immediate conflict of interest here, in that international golf regulations lay down clear duties with regard to the height of the grass and to the shaping of certain areas of the course. The *Professional Golfers Association* even prescribes how the greens are to be prepared months before any tournament.[50]

As early as the end of the 1980s it was pointed out to golf course architects and other experts in Germany, that 'a course whose greens are too narrow, … [would] have next to no chance of being considered for championships'.[51] Accordingly there are very few courses in Germany at the moment which could be considered suitable for national and regional championships. In 1996 the figure was given as 30.[52] Seen in the light of sport history, the concept of 'golf biotope networking' which has been co-promoted by the *German Golf Association,* reveals a remarkable development. Faced with the choice between ecology and competition, a sporting body has opted in favour of ecology. This not only puts in question the constitutive principle of sport, but is simultaneously an outright rejection of the sporting tradition of internationalism. For the international golfing scene has a clear preference for championship golf courses that have been designed along American

---

50 K.F. Grohs, 'Beeinträchtigen die funktionalen Anforderungen des Turniersports das erklärte Ziel "Landschaftsgolfplatz"?', in Internationale Vereinigung Sport- und Freizeiteinrichtungen e.V. (ed.), *Golftag '97.* Köln: IAKS 1998, 9–14, here 11.
51 Arbeitsgruppe 'Golfplätze' des Bundesinstitutes für Sportwissenschaften *(ed.), Planung, Bau und Unterhaltung von Golfplätzen* (Schriftenreihe Sport und Freizeitanlagen P1 / 1987), Köln: Verlag 1987, 15.
52 Grohs, 'Genehmigung', 14.

models.[53] That said, the members of the *German Golf Association* can still expect to enjoy international conditions on golf courses all over the world, but when at home they will have to make do with more down-to-earth solutions.

Seen thus, it cannot be an accident that it was a former Environment Minister from the Green Party – and not a representative of the *German Golf Association* – who reminded his readers in a recent glossy publication on the theme of 'biotope management on golf courses', that sports clubs must have 'the certainty that they will still be able to use their courses, even when their nature protection value has increased precisely because of this use'.[54]

---

53 W. Scheffler, 'Deutschlands schönste Golfplätze. Sind auch Design-Kriterien gefragt?', in *Golftag '97*, 31–35; D. Lémery, 'Golf course architect, between art and industry', in European Institute of Golf Course Architects, http://www.eigca.org/articles6.php (accessed 1 August 2006).
54 Quoted from: *Biotopmanagement*, 28.

NIKO LIPSANEN

12 Changing Places of Sport in the Light of Photographs: Helsinki 1976 & 2006

The focus of this chapter is on the changes in outdoor places of sport in Helsinki during the 30-year period from 1976 to 2006 as evidenced in photographs. The purpose is to try and discover what alterations have occurred to the buildings and situation of the sports sites and their layout along with developments in their vegetation and changes in possible usage. The photographs for the year 1976 were taken by Pentti Lagus for Helsinki City Sports Department and archived in Helsinki City Museum's Picture Archives. Only municipal sport sites are included in this collection, but photographs are available for several different kinds of outdoor sports grounds with a variety of facilities. The Lagus photographs have been matched by new photographs taken by the author in 2006, helped by a grant from Helsinki City. All the chosen sites were revisited and photographed again by the author in the summer of 2006. In addition, five new sport sites that did not exist in 1976 were photographed.

The photographs taken by Lagus included both summer and winter time shots. For this study only the summer photographs were included. While not denying the role of seasonality in the sportscape of Helsinki, winter photographs are less useful for our purposes because the landscape is often obscured by snow. The photographs show some kind of sports facilities in practically all the quarters and suburbs of the city. The most modest are just open graveled grounds with goals for soccer, but other grounds have a varied range of facilities, such as basket ball or tennis courts, adjacent to the main ground as well as seats for watching games, changing rooms or storehouses.

The photographic evidence

Primary sources for this study are two sets of photographs of sport sites, or places of sport, in Helsinki. The first set taken by Pentti Lagus in 1976 consists of more than hundred summer and winter photographs of municipal outdoor sport sites in different parts of Helsinki. From these 44 summer photographs of 28 different sites were selected for this study. They are black and white panoramic landscape photographs taken with a fisheye type of lens. The focus is on general views of the grounds and not on details.

According to Lagus, the idea for taking the photographs came from himself and the purpose was to have an inventory of the sites which would identify what kind of facilities they had.[1]

In 2006, the sites were visited and photographed again. Where possible, the spots where the original photographs were taken were located and the new ones taken in the same direction as in 1976. The idea was to reproduce the same landscape as carefully as possible, to enable comparison. The 2006 photographs were taken with Konica Minolta Z2 digital camera with a wide converter Minolta ZCW-100 enabling focal length equivalent to 28 mm in traditional 35 mm film photography. A tripod was used. The 28 mm equivalent focal length was not, however, wide enough to reproduce the views taken with a fisheye lens in 1976. For this reason, two photographs were taken side by side and combined with GIMP image manipulation software. After this operation, the resulting photographs were somewhat wider than the originals. In order not to lose any data, digital manipulation was kept to the minimum. The line between the two photographs is clearly visible in the final photographs. This may diminish the artistic value, if any, of the photographs but guarantees that no documentary data has been lost. The 2006 photoset was taken in colour. In addition to the panoramic landscape shots reproducing the views of 1976 photographs, more detailed photographs from and around the grounds were taken to gather additional information.

Methodological issues

Richard H. Hart and William A. Laycock cite the oft-quoted proverb 'one picture is worth a thousand words'. According to them, a 'pair or sequence of photographs, taken over time, can be even more valuable than a single photograph for documenting change in range or forest vegetation'. They emphasise that often photographs are the only data available for certain past conditions.[2] Research on vegetation change based on rephotography is extensive. James L. Zier and William L. Baker, for instance, systematically searched for historical photographs about the San Juan Mountains in Colorado, United States, and rephotographed them.[3] The method is not, of course, restricted to analysing vegetation changes but seems to be popular in documenting changes in geological or geomorphological conditions too.

Christian A. Kull has undertaken extensive work in Madagascar in rephotographing historical photographs of the landscape mostly taken by missionaries. He emphasises the problems in making conclusions based on insufficient data. For instance, in comparing two photographs of the same hill taken in years 1900 and 1996, one might conclude that there had been minor reforestation of this mainly grass-covered landscape feature. However,

---

1 P. Lagus, personal communication via telephone. June 12th, 2008.
2 R.H. Hart and W.A. Laycock, 'Repeat photography on range and forest lands in the western United States', *Journal of Range Management* 49: 1 (1996), 60.
3 J.L. Zier and W.L. Baker, 'A century of vegetation change in the San Juan Mountains, Colorado: an analysis using repeat photography', *Forest Ecology and Management* 228: 1–3 (2006), 251–262.

an intervening photograph from year 1946 of the same site shows that the hill was almost completely forested at the time. If that photograph had not been available the conclusion on the development of vegetation cover could have been misleading.[4]

Concerning photographs as evidence there is, according to Peter Burke, 'a continuing conflict between "positivists", who believe that images convey reliable information about the external world and the sceptics or structuralists who assert that they do not.'[5] The difference between them, according to Burke, is that while the positivists try to glimpse the real world beyond the image, structuralists draw attention to the internal organisation of the picture itself and its relation to other images in the same genre. Douglas Booth categorises the approaches in three ways: reconstructionism, constructionism and deconstructionalism. For the first approach, photographs are *prima facie* evidence; for the constructionalists they are 'contextualised within theory to reveal reality'. In deconstructionalism, photographs are 'ambiguous text passed through many mediating filters' and it is normal to have different interpretations of them.[6]

Burke emphasises the role of images as an important form of historical evidence while reminding us that their testimony also raises problems. Source criticism is an essential part of analysis when images are studied.[7] In analysing the changes of sportscape it is important to remember the limitations of the source material. What is visible in the photographs is not the whole nor necessarily even a representative picture of the situation. John Gilmore describes old postcards as 'glimpses of our past'.[8] The same is true about old photographs. They provide evidence of the past conditions but they do not tell the whole story. Their evidence is limited to how a particular scene framed by the photographer looked like at a particular moment in the past. It is possible to draw conclusions from the evidence but one has to keep the limitations in mind. One should also be aware of the known or possible purposes of photographers even in the case of documentary work. According to Burke, 'it would be unwise to attribute to these artist-reporters an "innocent eye" in the sense of gaze which is totally objective, free from expectations or prejudices of any kind', and stresses that one should not ignore the possibility of propaganda or stereotyped views.[9]

Nevertheless, making assumptions of a political or other kind of agenda seems not relevant in this case. The 1976 photographs were taken for a documentary or administrative purpose and they were all taken in a similar manner: panoramic landscape shot from a spot with good view over the

---

4 C.A. Kull, 'Historical landscape repeat photography as a tool for land use change research', *Norwegian Journal of Geography* 59 (2005), 258.
5 P. Burke, *Eyewitnessing: The uses of images as historical evidence*. London: Reaktion Books 2001, 184.
6 D. Booth, 'Evidence revisited: interpreting historical materials in sport history', *Rethinking History* 9: 4 (2005): 473.
7 Burke, *Eyewitnessing*, 14–15.
8 J. Gilmore, *Glimpses of Our Past: Social History of the Caribbean Postcards*. Kingston: Ian Randle 1995, ix.
9 Burke, *Eyewitnessing*, 19

grounds so that the grounds are seen in the photograph as completely as possible. If that was not possible, another photograph was taken to complete the documentation. In the case of grounds where changes since 1976 were few it was easy to find the original spots 30 year later as they were at most sites very obvious. It is relatively safe to assume that it has been the case also in the photographs where the original spot is no more the most obvious one due to growing vegetation or other changes.

Tapio Heikkilä has made an extensive work of visual monitoring of Finnish, mostly agricultural, landscapes based on systematic rephotographing. In Heikkilä's study, that means taking identically composed photographs 'at specified vantage points in a research site at certain intervals'.[10] Compared to this study Heikkilä's time span is considerably shorter, only a few years, but he has also taken the initial photographs himself. This may make the research more systematic than is the case here, where one is reproducing views taken decades before by another photographer in poorly documented conditions.

Again, the exact time when the photograph has been taken clearly matters. Fabrice Bensimon has noted about the photographs of Chartist demonstrations in London in 1848, that it is not possible to estimate the number of participants from the photographs since they were probably taken before the zenith of the event.[11] In this study for similar reasons it may not be possible to compare the volume of people using the sport sites in 1976 and 2006 on the basis of the photographs. While there are significantly more people in the 1976 photographs than those for 2006 it may be due to the timing. For the 1976 ones, the dates are not given, but as far as Lagus himself remembers, the 1976 photographs were mostly taken in the latter half of June.[12] In 2006, many of the photographs were taken in July and most of the rest in the first days of August, months when many people leave Helsinki for their holidays. The time of year is also important when we try to estimate the amount of vegetation. Early in the summer the leaves are smaller and vegetation in general is less lush. The day of the week and time of the day may also matter but the latter at least can be inferred from the direction of shadows (except on cloudy days). Most of the photographs both in 1976 and in 2006 were taken during the afternoon.

It is probably safe to assume that the people in the photographs of 1976 were in real life situations and not acting or posed. However, in 2006, when people at the grounds were very sparse, in a few cases I waited until someone came into the picture before taking it. Although the people are not acting, their appearance is not arbitrary either. It is possible that Lagus adopted a similar kind of practice in 1976 although there may not have been so much need as more people were obviously available.

10 T. Heikkilä, 'Visuaalinen maisemaseuranta: Teksti', *Taideteollisen korkeakoulun julkaisusarja* A 76.
11 F. Bensimon, 'Aux origines de la photographie historique: la première représentation d'une foule protestaire', in P. Dupuy (ed.), *Histoire, images, imaginaire*. Edizioni Plus: Pisa 2002, 141.
12 Lagus, personal communication.

Weather conditions at summers 1976 and 2006 seem to have been relatively similar at the time of taking the photographs. The weather is mostly sunny or partly clouded in the majority of 1976 photographs and in all the 2006 ones. In 1976, however, a few photographs were taken just after rain. This raises the issue of distinguishing the differences that are caused by the weather from those due to the course of time. This has to be taken into account when estimating the level of management of the grounds and the lawns around them.

Changes in the sportscape

In the title of this chapter I refer to sport sites as 'places of sport'. In most cases, these two concepts are interchangeable. Nevertheless, not all sites qualify as places since 'place', according to one dictionary entry, is a 'particular point on the earth's surface; an identifiable location for a situation imbued with human values'.[13] The sites studied in this chapter share, however, one important attribute of place which is name. Names are given to places and not, for instance, to landscapes.[14] Landscape, often understood as a panoramic view or scenery, is the physical form of place.[15] It is the landscape that is seen in the photographs of this study. In the context of places of sport it can be termed 'sportscape'.

The most common change in the sportscapes seen from the photographic comparison is a simple one: trees around the sites have grown bigger. Not only the trees but also the vegetation in general appears lusher in the 2006 photographs than in the 1976 ones. In many cases buildings that figure in the old photographs are barely visible in the later ones as they are hidden by trees. In a few instances it was no longer possible to reproduce the views taken by Lagus since most of the sports ground as seen from the original camera spot was not visible because of the vegetation.

It is arguable whether the lush vegetation is only a cause of time or if it reflects the less intense management of green areas currently in vogue (see Chapter Nine by Suvi Talja). Looking at the photographs of new sports grounds which did not exist in 1976 the first argument gets some support. Vegetation around new grounds is less lush than around the old ones pictured in 2006. From the photographs, it would seem that the immediate surroundings of sports grounds are cleared in the construction phase but not heavily managed afterwards. Hence, the older the grounds are, the lusher the surroundings. But the issue of vegetation as well as the changing nature of buildings and layout of sports grounds can be explored further by taking individual cases in different parts of the city.

In the central city Kaisaniemi Park close to the Central Railway Station is one of the oldest parks in Helsinki. Although it has undergone extensive

---

13 S. Mayhew, *Oxford Dictionary of Geography*. Oxford: Oxford University Press 1997, 327.
14 E.S. Casey, *Getting Back into Place: Towards a Renewed Understanding of the Place-World*. Bloomington: Indiana University Press 1993, 24.
15 E. Relph, *Place and Placelessness*. London: Pion 1976, 30.

Figure 12.1 Taivalsaari sports ground, Helsinki: an open field in 1976. (Photograph: P. Lagus, Helsinki City Museum).

Figure 12.2 Taivalsaari sports ground, in 2006. There is an indoor tennis hall in 2006. The site of the open-air tennis court on the left was not visible in the 1976 photograph which is taken somewhat more to the right. (Photograph by the author).

renovations in its history, the open sports field has remained almost unchanged for the last 30 years. Another old sports ground close to the city centre, Väinämöinen, is also almost as it was in 1976. The cabin has been renewed, a litter bin introduced and the skating rink is now there in summertime, but otherwise the changes are few. That is not, however, the case at nearby Taivalsaari sports ground which in 1976 had only an open field. Identifying the place 30 years later was challenging since the grounds are completely changed, the site now hosting an indoor sports hall and tennis courts (see Figures 12.1 and 12.2).

Most of the sport grounds in the mostly pre-war suburbs close to the city centre are also relatively unchanged. Only the trees have grown larger, though sometimes with other changes to the vegetation. This is most strikingly demonstrated in Kallio, for which older photographs are also available in the Helsinki City Museum archives. In the pictures taken in the 1930s only a few small trees and bushes frame the sports ground. In 1976 the trees have grown but bushes from the hill on the eastern side of the ground have been cleared. In 2006 large green trees almost hide some of the buildings but the hill is again covered by bushes and also with small trees. Close to Kallio, the sports ground in Vallila demonstrates similar development from 1976 to 2006.

Within a five kilometre radius from the city centre, there are three extensive sports parks: Eläintarha, Meilahti and Lauttasaari. They have all undergone only minor changes since 1976. Eläintarha is the oldest one and a historical venue for competitive sports. From 1976 to 2006, the main ground seems to be relatively unchanged. The asphalt surface is increased and the surrounding lawns are kept more carefully. Outside the ground itself, however, new constructions have been built and the location of a small wooden hut there in 1976 could not be identified in the 2006 study. Post-war Meilahti and Lauttasaari sports parks remain close to their 1976 character, though some facilities, such as lightning and fences, had been renewed, and the vegetation around the sports grounds had increased.

Further away from the centre, extensive sports parks were already established by 1976 as part of the general planning development of the Helsinki area from the 1960s noted by Suvi Talja in Chapter Three. Most notable are Pirkkola, Laajasalo, Pukinmäki, Kontula and Myllypuro. These have undergone somewhat bigger changes. At Pirkkola, which is located in the Central Park, the main grounds are almost unchanged. However, new indoor facilities have been built and other open spaces shown in the 1976 photographs are now covered by buildings. Laajasalo and Pukinmäki sports parks have been completely renovated and expanded and here it was not possible to identify the exact locations where the 1976 photographs were taken. At Kontula, a new changing room and spectator facilities have been built. On the other hand, at Myllypuro the location where the 1976 photograph was taken has remained relatively unchanged. Only the management of the grass lawns seems to be more careful than it was in 1976. However, outside the frame of the original photograph one finds the advent of facilities such as changing rooms obviously constructed after 1976. Many of the new facilities were clearly standardised. New changing rooms in Pukinmäki, Kontula and Myllypuro are replicas of each other, and the same kind of facility was also found at Puotila sports ground (called Vartiokylä in 1976).

In the suburbs the remaining sports grounds can be roughly divided in two categories: the semi-abandoned and the well-kept (the majority of the grounds). There are three sports grounds that can be considered as semi-abandoned: Torpparinmäki, Hirvitie and Annala. In Torpparinmäki, the reason is clear: a new, bigger sports ground has been constructed not far away. The old ground stills exist but is apparently less heavily used, as the vegetation has grown strongly. There are no buildings on site. Hirvitie sports

Changing Places of Sport

Figure 12.3 Upper sports ground, Annala, Helsinki. In 1976 this was more like a playground with benches, swing and sandbox. (Photograph: P. Lagus, Helsinki City Museum).

Figure 12.4 Upper sports ground, Annala, 2006. This looks much the same as in 1976, except for the removal of playground features. (Photograph by the author).

ground in Herttoniemi has not changed much since 1976. Although located between Itäväylä highway and apartment blocks it appears as if situated in the middle of a dense forest. Its lack of use may be because the Herttoniemi area has several sport facilities next to the schools.

Annala ground looks like it has been forgotten. Benches, swings and sandbox have been removed from the upper part of the grounds (see Figures 12.3 and 12.4). Graveled grounds have patches of grass growing at the edges.

191

The reason for its decreased use is not apparent, though it may reflect the wider shift away from organised sport in Finland, particularly among young people. As we have noted, declining use may contribute to increased vegetation though the photographic evidence is not detailed enough to demonstrate whether it also contributes to greater biodiversity. Laajasuo sports ground, which in 1976 looked like an ordinary suburb ground, was used for storing rocks and landmasses from a nearby construction site in the summer of 2006.

The rest of the sport grounds in the suburbs seem to be relatively well-managed. They include Lehtisaari, Munkkiniemi, Kivitorppa, Konala, Pohjois-Haaga, Pakila, Metsälä, Jakomäki, Vesala (Kontula North), Puotila (Vartiokylä), Roihuvuori and Kulosaari. Unlike the semi-abandoned ones, many of these grounds are located next to schools. They have not changed much in the last 30 years. Apart from the vegetation growth, the most notable developments since 1976 include the increased use of fences and more careful maintenance of the grass areas around the sports grounds. The latter is probably due to the increased use of machinery since the 1970s. At Jakomäki, a 'no dogs' sign which cannot be found in the 1976 photographs is visible in 2006.

For comparison, a few newer sports grounds that did not exist in 1976 were included in the study. These were Toukolanranta, Lassila, Torpparinmäki, Herttoniemenranta and Kallahti. These grounds show similar characteristics to those well-managed older grounds that have undergone some upgrading. They are mostly fenced, heavily managed and the sports grounds are clearly separated from the surroundings. The Herttoniemenranta ground is special due to its central location in the residential area. Here the fencing is essential because otherwise the ground would only be an open square in the midst of the buildings. That is also the case with some of the older sports grounds, most notably Vallilla and Munkkiniemi, which have not changed much since 1976. Of the newer grounds Toukolanranta is partly surrounded by residential buildings. By contrast those built in the suburbs are usually still surrounded by forest.

Overall, modest changes in buildings, layout and management seem to be characteristic of many of the grounds; Konala (see Figures 12.5 and 12.6) is a good example. One can clearly see the growing vegetation and the new fencing which does not, however, surround the whole ground (this is true at other sites too). At Konala, however, against the trend the management of the lawns is not apparently much more intensive in 2006 than in 1976. Of the new sports grounds, Toukolanranta is a good example of extensive fencing and space management.

In other respects, however, the photographic evidence raises more questions than it answers. Was the relative absence of people in 2006 the result of a difference in timing as far as taking the photographs were concerned, or did it reflect wider changes in the use of traditional outside sports areas, reflecting the shift away from organised sports which is documented elsewhere in this volume. Certainly given the much higher population levels in the Helsinki area in 2006 than 30 years earlier the potential number of users was greater. Again, how far did the changes to the

Figure 12.5 Konala, Helsinki, 1976: example of a typical sports ground in the suburbs. (Photograph: P. Lagus, Helsinki City Museum).

Figure 12.6 Konala sports ground, 2006: good illustration of the most common changes such as growing vegetation and increased fencing. Tennis courts on the left were fenced already in 1976. (Photograph by the author).

vegetation affect biodiversity. On the one hand, the increased height of trees and ground cover in some places, not least the semi-abandoned sites, may be seen as promoting biodiversity; on the other hand, the increased management of spaces including greater maintenance of grass areas may have had a more negative effect.

Location of the grounds

The analysis of the photographs suggests that it is important to see the location of sports grounds in relationship to the built environment – apartment buildings, schools and other facilities – and that this relationship is dependent on the age of the sports grounds. Here, however, the photographs alone are not sufficient evidence and a study of maps and on-site observation is required.

Older sites dating from the pre-war period and located mainly in the central city were in most cases open grounds in the midst of buildings. They were more like open squares that could be used for sports as well as other activities. The photographs taken in 1976, however, do not show other uses, except for some cases where pedestrians used the grounds as shortcut paths. In some cases, such as Vallila and Väinämöinen, that is the case still in 2006. The sport sites in the new suburbs built in 1960s and 1970s are different. Instead of being surrounded by buildings, their grounds are more often found either inside a patch of forest or at least having a forested area on one or more sides of the grounds. They may still serve as pedestrian shortcuts, however; the 1976 photographs show several cases where there were people passing by or through the grounds. In many cases, however, schools were built on the side of these grounds – the grounds were part of a planned complex. The grounds were open areas mostly without fences in 1976 but many of them had been fenced by 2006. The practice of fencing shows the growing trend to restrict use of the space specifically to sports.

The new sports grounds that did not yet exist in 1976 show even more clearly that they are meant exclusively for sporting purposes. The grounds are fenced and well managed. Pedestrian alleys are built around the grounds, on the other side of the fence. The same tendency is clearly seen at those sites which have been intensively renovated since 1976 and less clearly in some others. A small hint is the 'not for dogs' sign at Jakomäki. Locations outside the Helsinki built up area are still preferred although as we have already noted there are a couple of examples of more centrally located ones, Herttoniemenranta and Toukolanranta. All new sites are located either next to or at least close to public schools.

Conclusion

Even if not sufficient evidence to produce a comprehensive description of the evolution of sport sites in Helsinki in the past thirty years, our photographs from 1976 and 2006 are an important source. Their evidence is based on their documentary power. The photographs suggest that the changes in the sportscape of Helsinki in past 30 years have been relatively small. Many sport sites in 2006 look much the same as 30 years earlier. There have been some changes in the vegetation, though the ecological significance is not completely clear. There are more buildings and facilities on the site, probably reflecting the heavy municipal expenditure on sports and sports sites particularly in the 1980s. There may have been some changes in usage, though again the photographic evidence is somewhat opaque. Overall, however, when changes appear, the tendency is clear. Sport sites since the 1970 have become more exclusively reserved for sports. As places, the grounds are now more and more places of sport than places for everyone. The concept of 'place' in English comes from the Latin word meaning open square. An open square is thus the archetype of place, or at least one of them. Fences exclude people around the grounds and make them places reserved for those doing sport. On the other hand, fencing is one way of delineating place, and clear borders may increase the placeness, the place identity, of places for those who are inside them. In this respect, the growing vegetation around the grounds may also increase the feeling of enclosure and hence placeness.

EPILOGUE

EPILOGUE

JUSSI S. JAUHIAINEN

13 Green Space and Sport: Golf, Parkour and Other Post-Disciplinary Opportunities

Sport has a controversial relationship with green space, as many chapters in this book illustrate. Sport both affects and is affected by the physical environment and landscape.[1] Many sporting activities are performed in open green space. Often this sport-related green space has been artificially modified to fit the rules and regulations of various sports. Thus green space is promoted but also destroyed by sport. Activities using large areas of green space, for instance, golf and cricket, change the landscape and biodiversity of species and ecosystems, as Jarmo Saarikivi discusses in detail in Chapter Eight about golf courses. In this concluding chapter I want to return to some of the issues raised in the Introduction and explored in the case studies, as well as indicating some opportunities for further research.

Green is still the dominant colour in many sports, even in football, the most widely played team sport in the world. But what does green space mean in the urban environment? Urban space can be divided materially between 'grey space' that consists of hard, impermeable and sealed surfaces such as concrete and tarmac and 'green space' that is predominantly soft, unsealed and permeable surfaces such as soil, grass, shrubs and trees. Green space can be further divided between linear, semi-natural, functional and amenity green space, the latter including outdoor sport areas.[2] In addition, green space may be private, semi-public or public. Such classification helps us to appreciate the variety of green space in our cities and to consider the reciprocal relationship between green space and sport. Furthermore, one has to recognise the differences, challenges and opportunities of various types of green space for enhancing local biodiversity, as Philip James and Emma Gardner analyse in Chapter Ten, on privately owned sports-related green spaces. Such diversity requires a holistic analysis of urban green spaces and careful management of biotopes, as Jürgen Breuste discusses in Chapter Seven on Germany.

---

1 J.R. Bale, *Sports Geography*. London: Taylor and Francis 2003.
2 C. Swanwick, N. Dunnett and H. Woolley, 'Nature, role and value of green space in towns and cities: an overview', *Built Environment* 29 (2002):2, 97–98.

Social demand for urban green spaces increases as a result of rapid urbanisation.[3] Sport-related interventions create new public urban and green areas as well as various types of private sites. One can divide the rationale for the production of green space into supply and demand approaches. The supply approach aims at the conservation of high-quality natural and landscape values and relies on visual, ecological and spatial attributes of the existing natural environment. The demand approach considers the usage of open spaces to fulfil the needs of the target population. The areas are customised to take account of the size and demographic variables of population, their values and preferences, residential distribution and density.[4] There is often conflict between the supply and demand for green space.

In many European cities, the design of green space in and for sport has had an important physical and material impact on the urban environment over a long time. This history is documented in detail from the nineteenth century onwards by Pim Kooij in Chapter Four on the Netherlands and Fulvia Grandizio in Chapter Six on Turin in Italy. The actual site for sport itself might occupy a small area, such as a football pitch. However, much bigger sites are often constructed around this pitch. Today these may include the stadium, related leisure areas, car parks, and transport connections, as well as other services linked to a football match. Global mega-events related to sport are major urban and regional policy tools for regenerating urban areas.[5] However, many small everyday places of sport can remain unmodified and undeveloped for decades, although they are more fenced today as Niko Lipsanen observes in Chapter Twelve about outdoor sport pitches in Helsinki.

Green space becomes socially constructed through sport events. This can be illustrated through golf, an outdoor individual sport attracting enormous popularity in recent years. At the end of the 1990s, the worldwide number of golf courses was estimated to be 35,000, of which about 60 per cent were located in the United States. There the developers constructed on average one new golf course every day.[6] Today golf courses are designed and constructed everywhere regardless of climatic conditions: from the northernmost part of the inhabited world that is most of the year frozen, to the hottest and driest deserts and from sparsely populated countryside to the dense metropolis. A rough global estimate of the number of golf courses by 2010 is over 40,000. Issues of golf and the landscape have been studied for over a quarter of century[7] as well as the relationship to society and nature.[8]

---

3   J. Choumert and J. Salanié. 'Provision of urban green spaces: some insights from economics', *Landscape Research* 33 (2008):3, 331–345.
4   T. Maruani and I. Amit-Cohen, 'Open space planning modes: a review of approaches and methods', *Landscape and Urban Planning* 81 (2007), 4
5   S. Essex and B. Chalkley, 'Mega-sporting events in urban and regional policy: a history of the Winter Olympics', *Planning Perspectives* 19 (2004):2, 201–232.
6   M.R. Terman, 'Natural links: naturalistic golf courses as wildlife habitat', *Landscape and Urban Planning* 38 (1997), 183–197.
7   R. Adams and J. Rooney Jr., 'Condo Canyon: an examination of emerging golf landscapes in America' *North American Culture* 1 (1984): 211–221.
8   A. Pleumaron. 'Course and effect: golf tourism in Thailand', *The Ecologist* 22 (1992):3, 104–110.

Even today, golf is perceived mostly to be for elite groups in society and harmful for environment.[9] The boom in golf course construction creates conflicts between the interest groups of business development, wildlife protection and social inclusion, as Christiane Eisenberg and Reet Tamme explain regarding Germany in Chapter Eleven. The grass always seems to be greener on the other side.

The issue about green space and sport leads on to a very significant topic in the early twenty-first century city, namely that of public space. Sport has a particular relationship to public space. On the one hand, everyday sport – from jogging to specific free-running parkour and Nordic walking – is conducted in public space: in grey urban, in greenish semi-urban and in green rural areas. The presence of human bodies constructs green space culturally and makes it socially open. This important development took gradually place during the twentieth century in Helsinki, as Chapters Two and Three by Katri Lento and Suvi Talja indicate. The early exclusive and male-dominated character of outdoor sport sites was gradually transformed by sport-for-all policies. In everyday sport, a variety of people from different gender, class, culture, nationality and ethnicity intermingle peacefully. These interacting bodies democratise green space by their rhythmic, broken and discursive flows. As regards public space, the geographer Ash Amin states that 'through and beyond the consumption and leisure practices, the experience of public space remains one of sociability and social recognition and general acceptance of the codes of civic conduct and the benefits of access to collective public resources'.[10] Sport in green space has thus had a much more significant role than is normally recognised.

However, sport also privatises and consumes open space both through the staging of temporary events or the creation of more permanent, consumption-oriented and fenced areas. There is also sport-related resistance against this trend. One example is free direct running through built environment and parks – that is parkour. The growth of parkour not only illustrates the fragmentation of green space in our cities but also the shrinking amount of freely accessible public space. By the early twenty-first century sport has become a very important global consumption issue that includes material, social and symbolic dimensions. Professional sport is presented, represented and consumed through television, internet and on-site venues. Major sport events are followed by billions of people. Following sport and visiting various sport events have become a specific subculture, and event tourism often takes place in open and green space.[11] Such global sport is not only big business but also involves geopolitics and identity politics taking place in concrete sites and performed by real bodies, as Christiane Eisenberg discusses in Chapter Five on early twentieth century Germany.[12] The links

---

9 M. Green, 'Changing policy priorities for sport in England: the emergence of elite sport development as a key policy concern', *Leisure Studies* 23 (2004):4, 365–385.
10 A. Amin, 'Collective culture and public urban space', *City* 12 (2008):1, 7–8
11 B. Green and L. Chali, 'Sport tourism as the celebration of subculture', *Annals of Tourism Research* 25 (1998):2, 275–281.
12 See also J.R. Bale, 'Sport and national identity: a geographical view', *British Journal of Sports History* 3 (1986), 18–41; J. Praicheaux, 'For a geopolitical analysis of Olympic

between sport and nationalism have certainly not withered over time as the recent Olympic Games in Beijing illustrated. Of course global sport such as football also makes a contribution to peaceful relationships between countries by facilitating and promoting international communication.[13]

Sport in space is increasingly about visual aspects of the body. Traditionally, the well-trained body has been admired by artists and spectators. The body is the centrepiece of sport[14] and its athletic beauty.[15] Nowadays it is not anymore only about who wins but also who looks better in competition or when doing either professional or amateur sport. Part of the visual appearance in sport involves clothing, but increasingly the body itself displays with piercing and tattoos the identity, country of origin or even sponsors of a sportsperson. The link with fashion is intense in all kinds of sports: competitive professional, amateur, fun and health-related sport targeted to different segments of society. Even if one does not play sport, one is immersed in a sport-pervaded culture. Global sporting trademarks are present today everywhere and they have become part of everyday life of people, also outside the sport fields and activities.[16] Sport is about representation in which the original meaning has been overlaid by various commercial simulacra.

The visual aspects of sport also have a strong effect on the urban environment. Major sport arenas are designed by world-famous architects. Urban landscaping is used to make the site look better, often for the television spectators. However, much of the (professional) sport-related interventions into green space tend to be exclusive and tend not to offer inclusive landscape and recreation spaces for less wealthy and less powerful social groups, such as immigrants.[17] This is a challenge because lack of outdoor sport facilities contributes negatively to residential quality and attachment to neighbourhood. The quality of built-up space has a positive correlation with sports facilities and suitability of neighbourhood green areas.[18] Furthermore, there is a connection between deviant behaviour and the lack of green spaces or a lack of sports fields and playgrounds in a neighbourhood.[19] Environmental and policy measures such as separating buildings from parking lots by green

---

performance', *Sport Place International* 6 (1992):2, 27–34; H. Eichberg, *Body Cultures: Essays on Sport, Space and Identity*. London: Routledge 1998.

13 P. Hough, '"Make goals not war": the contribution of international football to world peace', *International Journal of the History of Sports* 25 (2008):10, 1287–1305.

14 Eichberg, *Body Cultures*.

15 C. Young, 'Kantian kin(a)esthetics: premises, problems and possibilities of Hans Ulrich Gumbrecht's in praise of athletic beauty', *Sport in History* 28 (2008):1, 5–25.

16 J. Mullin, S. Hardy and W. Sutton, *Sport Marketing*. New York: Human Kinetics Publishers 2007.

17 C. Risbeth, 'Ethnic minority groups and the design of public open space: an inclusive landscape?', *Landscape Research* 26 (2001):4, 351–366.

18 M. Bonaiuto, F. Fornara and M. Bonnes, 'Indexes of perceived residential environment quality and neighbourhood attachment in urban environments: a confirmation study on the city of Rome', in *Landscape and Urban Planning* 65 (2003) : 1–2, 41–52.

19 X. Bonnefoy, M. Braubach, B. Mossonnier, K. Monolbaev and N. Röbbel, 'Housing and health in Europe: preliminary results of a pan-European study', *American Journal of Public Health* 93 (2003):9, 1559–1563.

space help to promote positive physical activity by local people.[20] Walkable green environments in a neighbourhood encourage everyday mobility and contribute positively to the health and longevity of senior citizens in larger cities.[21] Nevertheless, one still finds arguments that the provision of urban green spaces should be based on economic rather than social criteria to ensure the efficiency of public spending.[22]

New opportunities in the study of green space and sport

To conclude, I would like to suggest some new research opportunities in the study of green space and sport. More inter-disciplinary approaches are needed, as Peter Clark, Marjaana Niemi and Jari Niemelä rightly point out in the Introductory Chapter. Unfortunately, academic disciplinary walls still surround the study of green space and sport. For example, a medical scholar researches outdoor sport and its relation to health;[23] a historian studies a sports event of the past in a green setting;[24] an economist counts the money flows related to major sporting events;[25] a political scientist observes the construction of a nation in the Olympic Games;[26] a gender scholar criticises the social construction of the female athlete body represented in the media;[27] a geographer describes the diffusion of various sports teams around the world;[28] a natural scientist analyses the variation of avifauna in golf courses of a particular region;[29] and so on. Too often disciplinary units constrain cooperation and common ground and inhibit fund raising for projects from an inter-disciplinary perspective.

Obviously, each academic discipline brings valuable information about particular topics. Without earlier discipline-based contributions, we would definitely have less in-depth knowledge about many aspects of green space and sport. Nevertheless, to follow the old practice of academic dissection is insufficient to address the complex, intertwined relationship of sport and green space. This book is an important first step towards an inter-disciplinary approach for the study of green space and sport. However, I would argue for

---

20 J. Sallis, A. Bauman and M. Pratt, 'Environmental and policy interventions to promote physical activity', *American Journal of Preventive Medicine* 15 (1998):4, 379–397.
21 T. Takano, K. Nakamura and M. Watanabe, 'Urban residential environments and senior citizens' longevity in megacity areas: the importance of walkable green spaces', *Journal of Epidemiology and Community Health* 56 (2002), 913–918.
22 Choumert and Salanié, 'Provision of urban green spaces'.
23 Takano, Nakamura and Watanabe, 'Urban residential environments'.
24 H. Eichberg, 'Race track and labyrinth: the space of physical culture in Berlin', *Journal of Sport History* 17 (1990):2, 245–260.
25 J. Siegfried and A. Zimbalist, 'The economics of sports facilities and their communities', *The Journal of Economic Perspectives* 14 (2000):3, 95–114.
26 Praicheaux, 'For a geopolitical analysis'.
27 M. Tervo. 'Geographies in the making: reflections on sports, the media, and national identity in Finland', *Nordia Geographical Publications* 32 (2003):1.
28 Bale, *Sports Geography*.
29 A. Sorace and M. Visentin, 'Avian diversity of golf courses and surrounding landscapes in Italy', *Landscape and Urban Planning* 81 (2006):1–2, 81–90.

even more intensive cooperation to address emerging themes in green space and sport. Instead of inter-disciplinary research, one could think about post-disciplinary approaches. The notion of a 'discipline' itself creates an unnecessary path-dependence in our approaches. I want to illustrate the opportunities of a more post-disciplinary approach with empirical cases from two sports gaining popularity, namely golf and parkour. Both have very particular trajectories in which humans and non-human participants meet in an interesting way.

As noted, there is a boom in constructing golf courses throughout the world. Traditionally, golf is often viewed from opposing perspectives. On the one hand, it is regarded as an elitist way of using time and space, destroying natural resources such as water and landscape, and creating exclusive areas for wealthy people. On the other hand, the players and followers of golf see it as good for the body and mind and helping to create beautiful human-made landscape areas.[30]

The 'greenness' of golf relates to contradictory discourses in relation to lawned areas and what type of interventions should be made in green space. The argument over what constitutes appropriate landscaping practices for green space provides a locus for bringing together, at a discursive level, the kinds of socio-cultural perspectives and practices that dominate places and locations in late capitalist society. Traditional 'lawn owners' (golf players) and 'ecological activists' (non-players) both perceive their discourse as natural. In a way green space is a site of identity politics defining social and economic status.[31]

Only fairly recently have golf course managers started to recognise the environmental impact of golf courses. In the 1990s, there was an emerging interest in natural habitats on golf courses that resulted in an increase in the amount of wildlife habitats; for example, if golf course designers built them in urban areas and on degraded landscapes such as landfills, quarries or eroded land. A naturalistic golf course with substantial amounts of native wildlife habitat can complement biological reserves, military reservations, greenbelts, parks, farms, backyards, and other units of the regional habitat mosaic. In addition, the large amount of habitat on naturalistic courses reduces water runoff, irrigation, and chemical inputs.[32] Coupled with the popularity of golf, naturally landscaped golf courses engage thousands of additional people in wildlife habitat preservation issues.

Increasingly, the golfing community is responsive to aesthetic and environmental concerns. In a case study of about one hundred golf courses in the United Kingdom, it was found that almost all golf course managers (90 per cent) considered golf courses important for wildlife and a majority (over 60 per cent) wanted to promote wildlife. However, formal management planning took place in only a minority (43 per cent) of the courses surveyed.

30 C.A. Gange, D.E. Lindsay and J.M. Schofield, 'The ecology of golf courses', *Biologist* 50 (2003), 63–68.
31 R. Feagan and M. Ripmeester, 'Reading private green space: competing geographic identities at the level of the lawn', *Philosophy and Geography* 4 (2001):1, 79–95.
32 M.R. Terman, 'Natural links: naturalistic golf courses as wildlife habitat', *Landscape and Urban Planning* 38 (1997), 183–197.

Conflicts occurred over how to manage wildlife alongside golf, often due to the demands of club members.[33] However, the greater responsiveness to environmental concerns does not mean that golf courses were open to everyone. In fact golf courses represent a case of private or semi-public environmental space. Even in Finland, where the law gives everyone the right to wander in green areas and enjoy them freely, private golf courses are a substantial exception, as Suvi Talja points out in Chapter Nine.

Golf courses are especially important for birds, as Jarmo Saarikivi stresses in Chapter Eight. Many golf courses equal the natural area in the rich diversity of bird species, though not in the incidence of specific kinds of birds. A case study of nine golf courses showed a positive relation between bird diversity and tree diversity for each habitat type.[34] Similarly, the proportion of forested area was positively correlated to the richness of species and to the number of species sensitive to forest fragmentation. Golf courses of any age can enhance the local biodiversity of an area by providing a greater variety of habitats than intensively managed agricultural areas, even if, with some exceptions,[35] they play a minor role in the conservation of specialist species.[36] Despite various challenges such as irrigation and the usage of fertilizers and pesticides for the maintenance of golf courses, they have also a positive contribution to local small-scale biodiversity, as noted in the Introductory Chapter. Inter-disciplinary study helps to highlight different aspects related to this particular sport in green space.

Let us go further and look at some of the opportunities for post-disciplinary perspectives in the study of green space and sport. One way forward is to approach it from alternative epistemologies and methodologies. An interesting issue is the utilisation of actor-network theory (ANT) for studying green space and sport. ANT has been an important approach to study complex heterogeneous relations and networks for over two decades.[37]

Particularly fruitful is the way ANT gets rid of canonised hierarchies and categories. The basic premise of ANT is that it starts from the middle of things through which relational practices and interactive networks arise, change and achieve durable forms.[38] ANT helps us find an engaged and motivated network, a forum of interacting actors. Actors define their interests, convince other entities, and get actively involved within a network.[39] The symmetrical illusions of interaction and society are replaced with an exchange of properties between human and non-humans. People, organisations, objects,

---

33 R. Hammond and M. Hudson, 'The management of golf courses for the conservation of biodiversity', *Landscape and Urban Planning* 83 (2007), 127–136.
34 R.A. Tanner and A.C. Gange, 'Effects of golf courses on local biodiversity', *Landscape and Urban Planning* 71 (2004), 137–146.
35 P.G. Rodewald, M.J. Santiago and A.D. Rodewald, 'Habitat use of breeding red-headed woodpeckers on golf courses in Ohio', *Wildlife Society Bulletin* 33 (2005):2.
36 Soracea and Visentin, 'Avian diversity of golf courses'.
37 J. Murdoch, 'Towards a geography of heterogeneous associations', *Progress in Human Geography* 21 (1997): 3.
38 A. Ivakhiv, 'Toward a multicultural ecology', *Organization and Environment* 15 (2002): 4, 391–393.
39 B. Latour, *Reassembling the Social: An Introduction to Actor-Network-Theory*. Oxford: Oxford University Press 2005.

technologies, and spaces are all concurrently brought together in the performance, for example, in the case of golf courses.[40] The processes of association and ordering between entangled people, things and heterogeneous elements are then studied.

There are a number of refreshing aspects in the ANT approach with regard to green space and sport. First, it shakes up the taken-for-granted categories and arguments in defining the actors and values in regard to the subject. To start with, in ANT nature is not separated from society but the common whole is called socio-nature. The relationship of time and space is not fixed but becomes more fluid.[41] Even non-human organisms can be attributed ontological status in the processes of environmental change, much like their human counterparts.[42] A biodiversity issue in a golf course becomes, for instance, an issue of politicisation.

Second, ANT takes into account the active role and intertwining of both human and non-human actors. Too often we put the non-human aspects into passive positions, hidden in the background. Both human and non-human actors enrol other human and non-human actors into heterogeneous assemblages or networks. What if we agree that in a golf course we are immersed in a very complex partnership of human and non-human actors? What if we take the role of measurement equipment, computers, internet group discussion channels, animals, plants, bacteria, pesticides and so on actively and seriously into account in defining what is a variety of biotopes in golf courses? Each minor material object in the golf course plays an active role and should be analysed and recognised along with the various human stakeholders.

Third, to follow the ANT approach one needs epistemological and methodological openness – too often lacking in studies of green space and sport. Through ANT, we get more by following rather than concentrating on defining and categorising circulations.[43] A researcher follows stakeholders (human and non-human actors) and their engagements, motivations and objectives within hybrid networks. An unusual moment helps to illuminate, delineate and explain the networks. Transactions between human and non-human actors generate traceable associations and render multiple relations visible. Thus networks become particularly visible when things break down and relations within the network do not work well.[44]

Unfortunately, many scholars consider post-disciplinary approaches such as ANT as a challenge and not as an opportunity. I would argue that we need ANT (ants) to grasp what golf-related biodiversity really is about. Traditional, even if joined-up, inter-disciplinary historical, geographical and

---

40 V. van der Duim, 'Tourismscapes: an actor-network perspective', *Annals of Tourism Research* 34 (2007):4, 961–976.
41 Latour, *Reassembling the Social*, 108
42 H. Perkins, 'Ecologies of actor-networks and (non)social labor within the urban political economies of nature', *Geoforum* 38 (2007):6, 1152–1162.
43 B. Latour, 'On recalling ANT', in J. Law and J. Hassard (eds) *Actor Network Theory and After*. Oxford: Blackwell 1999, 20.
44 J.S. Jauhiainen. 'Smoking fish – tracing representations and actor networks in the Baltic Sea region', *Folia Geographica / Geografiski Raksti* XIII (2007), 23–34.

ecological study of golf courses cannot tell us that. To push a bit further the idea of this approach, even a golf ball, a tee or a club makes a difference, or an injured knee of the world's most famous golfer. Only in this way can we see how green space and sport form a particular socio-nature.

Continuing this discussion of how human and non-human actors intertwine in simultaneously localising and globalising trajectories. Let us take as the example of the less popular but emerging sport of parkour – the art of movement. In the past twenty years, parkour has gained increasing popularity especially among young people around the world. For practitioners, the European urban environment is an exceptional playground for the exploration of discordant, counter-cultural practices against those architectures of control that are increasingly present in public space.[45]

Parkour consists of finding innovative, challenging and potentially dangerous ways to traverse the city landscape. It means to jump between buildings, over concrete walls, down stair wells and across fences. The activity is a way of using obstacles *en route* in order to perform jumps and acrobatics. It can be considered both leisure and reflection, a form of art or even akin to an eastern philosophy requiring discipline, self-improvement and interdependence.[46]

Ash Amin notes how 'public spaces are marked by multiple temporalities, ranging from the slow walk of some and the frenzied passage of others, to variations in opening and closing times, and the different temporalities of modernity, tradition, memory and transformation'.[47] A practitioner of parkour, called a *parkouriste* or *traceur,* selects a straight trajectory in a city and attempts to move as quickly and efficiently as possible along it, climbing or and jumping over various kinds of obstacles – from fences to cars – primarily using only the human body. Urban space becomes an experimental instrument. From one perspective, parkour can be considered as a specific, non-competitive, urban gymnastic exercise. Insurance companies and medical doctors define it as extreme sport with a high potential for injury.[48] It is also a mode to transgress the border between private and public space, closed and open space, by stressing its fundamental principles: movement and freedom. Through its spatial practice, parkour aims to reclaim what it means to be a human being actively engaged with the surrounding environment.

Passing through urban and green parks a *traceur* helps us to pay more attention to how our surrounding environment is organised and how passively felt objects have an active role in our relation to green and open space. Banal everyday surroundings suddenly acquire more in-depth significance. The seemingly random movement and assemblage of humans and non-

---

45 S. Shahani, *Parkour: The Evolution of the Disparate Tradition.* 2008. http://shawnshahani.wordpress.com/2008/07/15/parkour-a-european-disparate-tradition/ accessed 5 August 2008
46 C. McLean, S. Houshian and J. Pike, 'Paediatric fractures sustained in Parkour (free running)', *Injury, International Journal of Care Injured* 37 (2006):8, 795–797.
47 Amin, 'Collective culture', 12.
48 J. Miller and S. Demoiny, 'Parkour: a new extreme sport and a case study', *Journal of Foot and Ankle Surgery* 47 (2008):1, 63–65.

humans in public spaces may be seen guided by habit, purposeful orientation, and the instructions of objects and signs. It is fundamental for us to make sense of the space, our place within it and our way through it.[49] A *traceur* running, climbing and jumping through and over urban and green parks, indicates the active function of non-human objects in green space and how these grey and green spaces are very important and contested public spaces.

Through all types of everyday sports, from jogging to Nordic and traditional walking, we make public space and construct our urban environment. Despite the fact that mega events of sport often distract our attention, it is this daily use and consumption of everyday of green space and sport in that should increasingly involve us as scholars and as citizens.

---

49 Amin, 'Collective culture', 8.

List of Figures and Tables

Figures

Figure 1.1 The development of sports areas and green spaces in European cities **11**

Figure 2.1 Children playing in a sand pit in Helsinki in 1933/ Photograph: Toini Saloranta, Albumit auki **27**

Figure 2.2 The Ball Park sports field in Helsinki in 1947/ The Sports Museum of Finland **30**

Figure 2.3 Two female gymnacists by a lake in Tampere c. 1950/ Photograph: Jussi Kangas, Vapriikki Photo Archives **33**

Figure 2.4 The Olympic rowing stadium, Helsinki/ Photograph: Aarne Pietinen, Helsinki City Museum **36**

Figure 4.1 Maliebaan Utrecht 1713/ De Utrechtse Archieven **59**

Figure 4.2 Plan for the Amsterdam Forest, 1934/ Bosplan 1934: Catalogue of the exposition **68**

Figure 4.3 Plan for sports facilities in the Pex Forest, The Hague 1933/ Dienst Stedelijke Ontwikkeling The Hague **71**

Figure 4.4 Interland competition in 1913 in Zwolle/ Historisch Centrum Overijssel **74**

Figure 5.1 Hockey match in Berlin, c. late 1920s/ The Carl und Liselott Diem-Institut, Deutsche Sporthochschule Köln **80**

Figure 5.2 Tennis court in Berlin, mid-1920s/ The Carl und Liselott Diem-Institut, Deutsche Sporthochschule Köln **81**

Figure 5.3 Turnhalle Zehlendorf, a modern multi-sport gymnasium, c. 1960s/ Landesarchiv Berlin, Wolfgang Albrecht, Nr. 208829 **85**

Figure 5.4 Suburban sports facilities, c. early 1970s/ The Carl und Liselott Diem-Institut, Deutsche Sporthochschule Köln **86**

Figure 6.1 The FIAT mountain camp at Sauze d'Oulx/ Italian Archivio e Centro Storico Fiat **96**

Figure 6.2 The Dopolavoro built by FIAT, Corso Moncalieri, 1929/ Italian Archivio e Centro Storico Fiat **99**

Figure 6.3 The Mirafiori Dopolavoro project, 1937/ Italian Archivio e Centro Storico Fiat **100**

Figure 6.4 The FIAT swimming pool centre, Corso Moncalieri/ Italian Archivio e Centro Storico Fiat **101**

Figure 6.5 View from the hill of *Italia '61* Exposition/ Courtesy of the Historical Archives of the City of Turin (Miscellanea Lavori Pubblici 1099/1) **104**

Figure 7.1 Urban landscapes in Germany **115**

Figure 7.2 Location of outdoor sports in urban parks – Pestalozzi-Park in Halle/Saale **117**

Figure 7.3 Protected landscapes in the Landscape Development Plan of the city of Halle/Saale **119**

Figure 7.4 Habitat mapping in Germany **121**

Figure 7.5 Interdisciplinary Catalogue of Criteria **123**

Figure 9.1 Tali and other golf courses in the Helsinki metropolitan area in 2008 **147**

Figure 9.2 Land use change around the Tali golf course, 1979–2002 **157**

Figure 10.1 Number of new UK golf course developments built 1985–1998 **165**

Figure 10.2 Percentage of new UK golf course developments, by multi-national organisations, during the 1990s **165**

Figure 11.1 Public Golf Eastside Berlin/ Photograph: Reet Tamme **174**

Figure 11.2 Berliner Golfclub Stolper Heide/ Photograph: Reet Tamme **178**

Figure 12.1 Taivalsaari sports ground, Helsinki, 1976/ Photograph: Pentti Lagus, Helsinki City Museum **189**

Figure 12.2 Taivalsaari sports ground, Helsinki, 2006/ Photograph: Niko Lipsanen **189**

Figure 12.3 Upper sports ground, Annala, Helsinki,1976/ Photograph: Pentti Lagus, Helsinki City Museum **191**

Figure 12.4 Upper sports ground, Annala, Helsinki, 2006/ Photograph: Niko Lipsanen **191**

Figure 12.5 Konala sports ground, Helsinki, 1976/ Photograph: Pentti Lagus, Helsinki City Museum **193**

Figure 12.6 Konala sports ground, Helsinki, 2006/ Photograph: Niko Lipsanen **193**

Tables

Table 2.1 Five most popular public events in Helsinki Jan. 1926 – Oct. 1926 **29**

Table 4.1 Dutch sports organisations (dates of establishment) **62**

Table 9.1 Number of members in *Helsingin Golfklubi ry*, and all Finnish golf clubs, 1970–2005 **154**

Table 10.1 Land use in Salford **163**

Notes on Contributors

Jürgen H. Breuste is Professor of Urban and Landscape Ecology in the Department of Geography and Geology at the University of Salzburg, Austria. He is the author of numerous articles and books on urban ecology and urban ecosystem management, and is president of the International Society of the Science of Urban Ecosystems (ISSUE) and of the IALE Landscape Research Centre (CeLaRe) in Germany.

Peter Clark is Professor of European Urban History at the University of Helsinki. In 2006 he edited *The European City and Green Space: London, Stockholm, Helsinki and St Petersburg 1850–2000* (Ashgate) and his *European Cities and Towns 400–2000* was recently published by Oxford University Press. He is currently coordinating a new international project on global cities from ancient times to the present.

Christiane Eisenberg is Professor for British History at the Centre for British Studies, Humboldt-Universität zu Berlin. Among her book publications are *"English sports" und deutsche Bürger: Eine Gesellschaftsgeschichte 1800–1939* (Schöningh 1999) and (with Pierre Lanfranchi, Tony Mason und Alfred Wahl) *FIFA 1904–2004: Le siècle du football, Paris* (Cherche Midi 2004), translated into German, English, Spanish and other languages.

Emma L. Gardner is an Honorary Research Fellow in the School of Environment and Life Sciences at the University of Salford, UK. Her doctoral studies examined the sustainable development issues associated with the development of a major hotel and golf club complex in Salford. She now works as a sustainability consultant for AECOM.

Fulvia Grandizio is an architect. She was awarded a PhD in the History of Architecture and History of Urbanism at the Faculty of Architecture, Polytechnic of Turin. Her fields of interest are urban history in the 19th and 20th centuries and the history and preservation of gardens. She has written articles on urban history and landscape architecture. At present she is collaborating on a survey of historical gardens of Regione Piemonte (Piedmont).

Philip James is a Reader in the School of Environment and Life Sciences at the University of Salford, UK. He has published extensively on issues related to social-ecological systems with particular reference to urban areas.

Jussi S. Jauhiainen is Professor of Geography at the University of Turku and Associate Professor of Urban Geography at the University of Tartu. He is the author of numerous articles on urban development, planning and policy.

Pim Kooij is Professor of Economic and Social History at the University of Groningen and Professor of Rural History at the University of Wageningen. His specialties are urban and rural history; he has written several books on Dutch cities, and on regional history, in which he combines urban and rural approaches.

Katri Lento is writing her PhD dissertation at the History Department, University of Helsinki. The dissertation analyses the roles and meanings of urban green space in Helsinki from the late 19th century to the 1960s.

Niko Lipsanen is a doctoral student in the Geography Department of the University of Helsinki. His thesis explores different aspects of the use and perception of green space in present day Helsinki.

Jari Niemelä is Professor of Urban Ecology at the Department of Environmental Sciences at the University of Helsinki. He is currently editing a book on urban ecology for Oxford University Press and has published a number of articles on urban ecology and related issues.

Marjaana Niemi is an Academy Research Fellow at the Department of History and Philosophy, University of Tampere. She is the author of *Public Health and Municipal Policy Making: Britain and Sweden* (Ashgate 2007), as well as numerous articles on urban history and the social history of medicine.

Jarmo Saarikivi is a doctoral student at the Department of Environmental Sciences, University of Helsinki. His research focuses on urban ecology and biodiversity. He has a particular interest in animal species assemblages in urban green space, especially sports sites such as golf courses.

Suvi Talja is a doctoral student at the History Department of the University of Helsinki. Her dissertation explores urban space and physical culture in Helsinki and Dublin between the 1950s and the early 21st century.

Reet Tamme is a historian; she is a Research Fellow at the Collaborative Research Center "Changing Representations of Social Order" at the Humboldt University Berlin. Her research project deals with the urban sociology of race relations in Britain from the 1950s to the 1970s.

Index

actor-network theory 205
aerobics 48, 49, 55, 105
Agnelli, Giovanni 97, 98, 100
agriculture 21, 42, 46, 51, 99, 103, 107–108, 114, 117, 125, 131, 146, 148, 160, 162–163, 166, 175, 177, 179, 187, 205
allotment gardens 10, 67, 124, 127
Amin, Ash 201, 207
amphibians 135, 136–137
Amsterdam 17, 59, 60–63, 65–68, 70, 72–75
   Sarphati Park 66
   Vondelpark 60, 66, 72
architects 16, 37, 69, 104, 105, 140, 171, 180, 182, 202
athletes 30, 32, 78, 79, 88, 95, 98
athletics 13, 27, 30, 31–34, 39, 45, 62–63, 67, 72–73, 79, 80, 84, 92, 97, 100–101, 172–173
Austria 9
baseball 33, 62
basketball 75, 97
Bavaria 119, 175
Belgium 74, 125
Berlin 21, 67, 80–81, 83, 85, 113, 117–119, 122, 172, 174–178
billiards 61–62
biodiversity 10, 18–19, 21, 105, 109, 126–127, 129, 131–133, 139, 142–143, 160, 163, 169, 170, 179, 192–193, 199, 205–206
birds 104, 107, 132–135, 137, 140, 205
black balling 63
boccie 93, 94, 98
bodily experimentation 78
body 25–26, 51, 78–79, 92, 94, 201–204, 207
bowling 28
Brandenburg 21, 172, 175–178
Breda 58, 60

Britain 9, 11, 12–14, 60, 64, 76–79, 81, 88, 131, 161–170, 172–173, 175
cafes and restaurants 26, 36, 59, 62, 73, 81, 93, 150
camping and camp sites 36, 38, 70, 94, 96
cemeteries and churchyards 10, 46, 58, 70, 127
city councils 26, 34–35, 54–55, 60, 64, 72, 83, 91, 102, 149, 152, 167
   Helsinki City Council 34–35, 55
   Salford City Council 167
   Turin City Council 102
city government 10, 15, 16, 17, 27, 43
city images 17
class and sports 13–14, 25–29, 37, 39, 62–66, 77, 81–82, 91–94, 106
Cold War 20, 85, 86
Collan, Anni 32
communism 113
contested space 9, 14, 21, 25, 39, 143, 208
convent gardens 58, 62
countryside 31, 35, 42, 60, 62, 75–76, 108–109, 125, 162, 164, 175, 177, 200
cricket 13, 62, 66–67, 70, 72, 76, 80, 93, 199
   pitches 13, 76
cycling 28, 62–63, 65, 67, 73, 77, 80
Delft 61
   Agneta Park 61
developers 16, 20, 69, 145, 200
Dokkum 62
East Germany, see German Democratic Republic
ecology and ecologists 9, 10, 18–19, 21–22, 106, 113–114, 126, 131, 133, 141–142, 144, 161, 164, 166, 182
ecological
   framework 21, 161, 169
   golf business 182
   processes 9, 161

214

resources 9, 18, 128
    significance of sports sites 9, 17–21, 107, 128, 194
ecosystems 19–20, 113, 118, 127, 133, 141–142, 161, 199
Eindhoven 61, 65
engineers 76, 94, 104
England, 63, 92, 130, 145, 170, *see also* Britain
English landscape style 67
environmental
    education 114, 121–122
    organisations 22, 105, 109, 172
    problems 21–22, 124, 144
Erkko, Eljas 154
ethnic minorities 11, 17
eugenics 13
European Golf Association 106–107, 144
European Union 105, 109, 122
exhibitions and expositions 28, 92–93, 96, 103–104
extension plans 66, 74–75
factory sports grounds 13, 84, 98–101
fascism 95, 97, 102
Finnish Athletics Association (SVUL) 30–32, 38–39
Finnish Workers' Sports Federation (TUL) 30–31, 38–39
Federal Republic of Germany 20–21, 113, 116, 118, 120, 124, 173, 175–176
female exercise 32–34
femininity 20, 32, 57
fencing 156, 192–195
fencing (sport) 61–62, 67, 73, 93
FIAT 20, 91, 96–102, 107, 109
Finland 9, 11, 14, 19, 21, 25–57, 135, 143–160, 192, 205
    and Russification 19, 29
First World War 9, 13, 15, 18, 20, 63, 72, 79, 82–83, 95, 98, 128
fives 61–63
football 11, 13, 33, 51, 55–56, 62–63, 66–67, 70, 72–73, 75–77, 93, 96, 98, 150–151, 172–173, 199, 202
    pitches 14, 50–51, 64–67, 70, 72, 75, 80, 84, 92, 97, 144, 151, 200
forests and woodlands 18, 20, 38, 46, 47, 49, 51, 58, 61, 75, 119–120, 128, 132, 134, 137, 144, 151, 157, 164, 167–168
fortifications 60–61, 65, 71, 73, 75, 91
Franeker 62
garden city concept 61
garden suburbs 25, 64–65
gardens 10, 27, 58–59, 67, 91, 92–93, 102, 124, 127, 140, 163, 167–169
gardeners 61

gender and sports 16, 19–20, 31–33, 37, 39, 53–55, 57, 203
German Democratic Republic 21, 87, 116, 124, 172, 176–177
German Golf Association 22, 173, 180–183
German reunification 1989/90 87, 172
Germany 9, 12, 14, 20–22, 33, 64–65, 76–89, 113–124, 127, 145, 171–183, 199, 201
global changes 17, 77, 125–126, 135, 171–172, 181, 200–202, 207
Golden Plan for Health, Sport and Recreation 84
golf 13, 62, 80, 88, 105–106, 109, 116, 199–200, 204
    image of 10, 106, 116, 143–144, 171–173, 201
golf boom 10, 14, 21, 130, 171–183, 200, 204
    in Britain 164
    in Finland 48, 145, 155
    in Germany 21–22, 88, 171, 201
golf courses 10, 14, 17–22, 44, 48, 75, 88, 91, 106–108, 126–160, 164–183, 199–200, 203–207
    environmental impact of 10, 17–19, 22, 107, 126, 130–141, 143, 156–161, 166–170, 179– 183, 199–100, 203–207
    maintenance of 19, 21–22, 107, 126, 129, 136–140, 156–160, 166–170, 204– 205
grass sports 79–80
Greater Manchester 163–164
green belt 66, 72–73, 75, 83, 162,
green city 113
green corridors 45, 57, 68, 105, 117, 124, 133, 137, 140, 168–170
grey space 14, 82, 84, 86, 88–89, 97, 199, 201, 208
Groningen 58–59, 65, 71–75
gymnasiums 34, 81, 83, 85, 93, 98, 101
gymnastics 31–34, 48–49, 62–63, 72, 78–79, 82, 92, 173,
Haarlem 58, 63–64
habitat mapping 118, 120–121
habitats 18–19, 118, 120–121, 126–142, 144, 157, 164, 168–169, 180, 204–205
the Hague 58–59, 61–63, 65, 69–71, 75
Hamburg 83, 175
handball 62–63, 67, 72, 84
hedgerows 167–168
Helsingin Golfiklubi ry 145–148, 151–160
Helsingin Sanomat 47, 51, 55, 151, 154
Helsinki 13–14, 16–22, 25–57, 75, 125–

127, 135, 137, 139, 144, 146–160, 184, 187–194, 200–201
  Ball park (pallokenttä) 30, 36
  Central park 33, 36, 50, 190
  Eläintarha sports park 33, 35, 190
  Kaisaniemi park 28, 49, 188
  Master Plan 44–45, 150, 152–153, 156
  Myllypuro suburb 45, 47, 50–52, 190,
  Pirkkola sports park 45, 47, 50, 151, 190
  Tali golf course 18, 21, 143, 145–147, 149–160
Hengelo 61, 65
hockey 13, 53–54, 63, 67, 79–80
horse riding 28, 61, 65, 67, 105
House of Savoy 90
housing 25, 36, 44–46, 50, 60, 65, 74, 103–104, 125, 140, 148, 150, 152, 156, 160, 162, 164, 167–168, 173, 176, 179
hunting 58, 61–62, 90, 107–108, 125
identity
  national 12–13, 19, 37
  place 22, 195
images, national 39
indoor sports sites 12–14, 20, 45, 47–48, 50–51, 53, 56–57, 75, 81, 84, 97, 100–101, 189, 190
industrial areas 114
industrialisation 10, 58, 60–61, 76
industries 46, 162–163, 175
insects 132, 140
invertebrates 132, 138
Italy 9, 14, 90–109, 200
jogging 14, 46–47, 51–52, 75, 105, 156, 201, 208
Jung, Bertel 33
korfball 62–63, 65, 67, 72
Lagus, Pentti 184–185, 187–189, 191, 193
land use 12, 94, 113–114, 118–120, 124, 128, 139, 152, 157, 160–161, 163, 169
landscape 9, 16–19, 21–22, 39, 41, 60–61, 67, 71–73, 83–84, 103, 105–108, 113–115, 117–120, 128, 131–133, 136–137, 139–141, 143, 146, 153, 161, 167, 169, 171, 177, 180, 184–188, 199–200, 202, 204, 207
  ecology 131, 161, 164, 166
language groups 16, 19, 30
legislation
  Finnish 27, 43–44, 52, 55–56
  Dutch 65, 74
  German 118, 175, 179, 181
  Italian 93, 102, 104
leisure
  time 13, 37–38, 41–42, 164
  activities 10, 52, 76, 79, 93, 98, 128, 129, 153

Leeuwarden 59
Leiden 59
Leipzig 122
London 9, 13, 16, 67, 187
  County Council 13
Maastricht 58
Maliebaan 59, 69
mammals 137–138, 180
masculinity 13, 20, 54, 57
mass movements 78
media 14, 31, 48, 51, 53, 78, 102, 151, 154, 171, 203, *see also* newspapers
mega-events 200, *see also* Olympic Games
memories 51
middle classes 13, 25, 50, 60, 62–66, 77–79, 81–82, 92, 94
military
  agendas 16, 77–78
  education 13, 63, 78
motor sports 35, 78
multi-functional spaces 114, 122, 128, 151
municipal
  authorities 15–16, 19, 22, 46, 52, 65, 90, 92, 160, 177, 181
  policies 9, 11, 14–15, 20, 22, 26, 67
Munich 86, 127, 175
Mussolini, Benito 94, 97,
mutual aid societies 99–100
national prestige 16–17, 30, 79
National Trust 60
natural environment 10, 46, 105, 114, 120, 125–126, 162, 166, 172, 200
nature 9–10, 13, 16, 21, 25, 56–57, 60–61, 67, 72, 80, 102, 105, 113–114, 118–122, 125–133, 137, 141–142, 144, 170, 200
  as an educational resource 121–122
  conservation 21, 60, 115, 118, 131–132, 169
  protection 21–22, 107, 113, 118–121, 171–172, 177, 179–183
  reserves 9, 18–19, 105, 128–129
Netherlands 9, 13–14, 58–75, 173, 200
network of sports sites 43–45
Nijmegen 58
Nordic countries 9, 50, 145
Northern America 32, 130, 133, 164, 185, 200, *see also* United States
Norway 134, 145
Obermann, Rudolf 92
Olsson, Svante 33–34, 36
Olympic Games 13, 17, 34, 37, 79, 203
  in Amsterdam 65–66
  in Athens 17
  in Beijing 202
  in Helsinki 35, 49, 148

in Stockholm 30
Olympic Stadium Berlin 83
outdoor sports sites 12–14, 32–33, 39, 45–47, 54, 72, 80, 90, 97, 105, 109, 116, 184, 199, 201–202
pall mall 59, 61–62
parkour 11, 199, 201, 204, 207
parks 12–13, 19–20, 26–28, 31, 33, 36, 38, 49, 61, 65–69, 72, 91–95, 103, 105, 107–108, 117, 128–129, 143, 149, 160, 166, 179, 188, 190
    municipal 12–13, 18, 26, 88, 109, 116
    royal 60
    villa 61, 69–70
people's parks 38, 46–47, 64, 149
periurban 113–114, 125, 164, 166
physical education 32, 34
placeness 22, 195
planners 16, 19, 21, 37, 42, 85, 87, 91, 105, 142, 150, 169
planning 9–10, 14, 16, 19–20, 26–27, 38–39, 43–45, 50, 53–54, 65, 74, 91, 102–103, 105–106, 109, 113, 115–117, 122–124, 129, 150–152, 156, 160, 167, 179, 190, 204
Planning Department, Helsinki 150, 156
plants 21, 118, 120–121, 126–127, 131–132, 140, 156, 167–169, 180
playground movement 32
playgrounds 26–27, 32, 34, 65, 102, 127, 202
playing fields 18–19, 43–44, 76–77, 81, 83, 95, 129, 132
photographic evidence 184–185, 192, 195
Po river 90, 93, 105
political ecology 9–10, 22
ponds 18, 63, 67, 72, 129, 132, 134, 136–137, 168, 180
private spaces 90, 170, 205, 207
public health 9, 15, 17, 38, 40, 78, 94
public spaces 11, 201, 207–208
racing, 62, 72, 93
recreation routes 41, 43–45, 47, 50, 55, 57, 151, 155
recreational
    areas 10, 32, 47, 67, 109, 128, 131, 177, 182
    needs 41, 114, 179
regeneration 162, 164, 167–168
regionality 9
rephotography 185
residential areas 47, 56, 88, 101, 114, 116, 124, 146, 150, 152, 155–156, 160, 192
residents, local 14, 146, 155–156
Rotterdam 61–62, 65
rowing 28, 36, 62, 67, 72, 93–94, 98

rugby 13, 63, 93,
sailing 28, 62, 65
Salford 21, 161–170
    Marriott Worsley Park 161, 166–170
school sports grounds 13, 32, 35
Scotland 108, 132
Second World War 16, 20, 37, 68, 70, 73, 84–85, 92, 95, 102, 104, 109, 113, 116, 124, 148
segregation 11, 33, 58, 63, 107
shooting 62, 73, 93, 172
skating 14, 28, 39, 46, 50–52, 56, 62–63, 67, 72–73, 81, 93, 117, 189
skiing
    cross-country 48
    down-hill 48
    tracks 38, 44, 51
    tunnels 56
skittles 61–62
social
    exclusion 171
    problems 38, 52–53,
space management 21, 124, 192
'Sport for all' 40–43, 50, 52, 54–55, 57, 201
sports
    amateur 12–14, 41, 202
    clubs and societies 14, 16, 28, 30–31, 37–39, 49, 55, 58, 63, 94, 181, 183
    commercial 11–14, 48, 56
    commercialisation of 42, 48–49, 57, 116
    competitive 11–12, 20, 30, 32, 37–39, 41–42, 44, 49, 52, 53, 56, 62, 85, 128, 190
    culture 41–42, 48, 55, 61,
    individualistic 12, 14
    indoor 42, 53, 56–57, 82, 93, 105,
    international 12–13, 20, 39
    mass 20, 28, 39–41, 44, 93, 102
    organisations 15, 19, 32, 35, 37, 39, 42–43, 52–43, 62, 82, 94
    organised 10, 12–14, 20, 55, 61–67, 72, 87, 89, 128, 192
    outdoor 13, 28, 33, 57, 102, 105, 109, 200, 203
    parks 33, 36, 44–45, 50–51, 64, 74, 83–86, 151–152, 190
sports departments
    Helsinki 38–39, 43, 48–49, 51, 156, 184
    Turin 102–103
sportsmen 11, 14, 38, 42, 63–64, 72, 76–78, 80
sportswomen 43
squash-halls 48

stadia 12–14, 36, 50, 53, 65–66, 69, 71–75, 79–80, 83–84, 92, 96–97, 100, 200
Stadtumbau Ost-Programm 176
state
    policies 20
    subsidies 43, 49, 65, 75
Stockholm 16, 30, 50
suburbs 25, 38, 42, 44, 49–55, 86, 101, 103–105, 108, 150, 160, 162–164, 166, 176, 184, 190, 192–194
Sukopp, Herbert 113, 119
surveys 47–48, 53, 102–103, 105, 107, 120, 130, 146, 164, 180
sustainability 9, 132, 143, 164
Sweden 34, 38, 145–146, 173, 175, 181
swimming 34, 41, 45–46, 62–63, 67,
    pools 13, 38, 46–47, 50–53, 65, 67, 70, 72, 75, 92, 94–95, 97, 101
tennis 13, 62–63, 66, 72–73
    courts 13, 36, 64, 67, 70, 81, 184, 189, 193
Treaty of Versailles 83
trees 59, 69, 95, 128–129, 134–135, 158, 161, 168, 180, 188, 190, 193, 199, 205
tourism and tourists 76, 106, 166, 171, 175, 177, 201
tuberculosis 15, 94, 98
Turin 20, 90–105, 108–109, 200
    the Hills 90, 95, 103
Turnen 78, 81–83
United States 130, 138, 164, 181, 185, 200
urban
    conflicts 16–17, 115
    culture 34–36
    history 10
    marketing 17
    sprawl 113, 125, 141, 162, 176
urbanisation 10, 13, 16, 41, 58, 60, 64, 76, 79, 103, 109, 125, 127, 131, 134, 136, 141–142, 150, 200
Utrecht 59–62, 65, 71
vegetation 22, 120, 135, 138, 140, 169, 184–188, 190, 192–195
velodromes 53, 96
Volkspark (people's park) 38, 46, 64–65, 70, 72
walking 47, 52, 62, 64, 66, 98, 201, 208
Washington 134
welfare
    policies 16, 38, 40, 43, 98–99, 109, 130
    state 38, 42–43
West Germany, *see* Federal Republic of Germany
wetlands 18, 119–120, 132, 134, 136–137, 141, 168
wildlife 18–19, 105, 107, 127–134, 137–138, 140–141, 164, 168–170, 201, 204–205
Wilskman, Ivar 26, 31
women 11, 13–14, 17, 31–34, 37–39, 43, 48–49, 51, 53–55, 62–64, 87, 93, 153
working classes 13–14, 26, 28, 30,–31, 39, 50, 61–65, 72–73, 77, 84, 91–92, 94, 98–100, 103–104, 109, 116, 162
Zwolle 65, 73–74

STUDIA FENNICA ETHNOLOGICA

*Making and Breaking of Borders
Ethnological Interpretations,
Presentations, Representations*
Edited by Teppo Korhonen,
Helena Ruotsala & Eeva Uusitalo
Studia Fennica Ethnologica 7
2003

*Memories of My Town
The Identities of Town Dwellers and
Their Places in Three Finnish Towns*
Edited by Anna-Maria Åström,
Pirjo Korkiakangas & Pia Olsson
Studia Fennica Ethnologica 8
2004

Passages Westward
Edited by Maria Lähteenmäki
& Hanna Snellman
Studia Fennica Ethnologica 9
2006

*Defining Self
Essays on emergent identities in Russia
Seventeenth to Nineteenth Centuries*
Edited by Michael Branch
Studia Fennica Ethnologica 10
2009

*Touching Things
Ethnological Aspects of Modern
Material Culture*
Edited by Pirjo Korkiakangas,
Tiina-Riitta Lappi & Heli Niskanen
Studia Fennica Ethnologica 11
2009

Gendered Rural Spaces
Edited by Pia Olsson & Helena Ruotsala
Studia Fennica Ethnologica 12
2009

STUDIA FENNICA FOLKLORISTICA

*Creating Diversities
Folklore, Religion and the Politics
of Heritage*
Edited by Anna-Leena Siikala,
Barbro Klein & Stein R. Mathisen
Studia Fennica Folkloristica 14
2004

Pertti J. Anttonen
*Tradition through Modernity
Postmodernism and the Nation-State
in Folklore Scholarship*
Studia Fennica Folkloristica 15
2005

*Narrating, Doing, Experiencing
Nordic Folkloristic Perspectives*
Edited by Annikki Kaivola-Bregenhøj,
Barbro Klein & Ulf Palmenfelt
Studia Fennica Folkloristica 16
2006

Mícheál Briody
*The Irish Folklore Commission
1935–1970
History, ideology, methodology*
Studia Fennica Folkloristica 17
2007

STUDIA FENNICA HISTORICA

*Medieval History Writing and Crusading
Ideology*
Edited by Tuomas M. S. Lehtonen
& Kurt Villads Jensen with
Janne Malkki and Katja Ritari
Studia Fennica Historica 9
2005

*Moving in the USSR
Western anomalies and Northern
wilderness*
Edited by Pekka Hakamies
Studia Fennica Historica 10
2005

Derek Fewster
*Visions of Past Glory
Nationalism and the Construction
of Early Finnish History*
Studia Fennica Historica 11
2006

Modernisation in Russia since 1900
Edited by Markku Kangaspuro
& Jeremy Smith
Studia Fennica Historica 12
2006

Seija-Riitta Laakso
Across the Oceans
Development of Overseas Business
Information Transmission 1815–1875
Studia Gennica Historica 13
2007

Industry and Modernism
Companies, Architecture and Identity
in the Nordic and Baltic Countries
during the High-Industrial Period
Edited by Anja Kervanto Nevanlinna
Studia Fennica Historica 14
2007

Charlotta Wolff
Noble conceptions of politics
in eighteenth-century Sweden
(ca 1740–1790)
Studia Fennica Historica 15
2008

Sport, Recreation and Green Space
in the European City
Edited by Peter Clarck, Marjaana
Niemi & Jari Niemelä
Studia Fennica Historica 16
2009

STUDIA FENNICA LINGUISTICA

Minna Saarelma-Maunumaa
Edhina Ekogidho – Names as Links
The Encounter between African and
European Anthroponymic Systems
among the Ambo People in Namibia
Studia Fennica Linguistica 11
2003

Minimal reference
The use of pronouns in Finnish
and Estonian discourse
Edited by Ritva Laury
Studia Fennica Linguistica 12
2005

Antti Leino
On Toponymic Constructions
as an Alternative to Naming Patterns
in Describing Finnish Lake Names
Studia Fennica Linguistica 13
2007

Talk in interaction
Comparative dimensions
Edited by Markku Haakana,
Minna Laakso & Jan Lindström
Studia Fennica Linguistica 14
2009

STUDIA FENNICA LITTERARIA

Changing Scenes
Encounters between European
and Finnish Fin de Siècle
Edited by Pirjo Lyytikäinen
Studia Fennica Litteraria 1
2003

Women's Voices
Female Authors and Feminist Criticism
in the Finnish Literary Tradition
Edited by Lea Rojola & Päivi
Lappalainen
Studia Fennica Litteraria 2
2007

Metaliterary Layers in Finnish Literature
Edited by Samuli Hägg, Erkki Sevänen
& Risto Turunen
Studia Fennica Litteraria 3
2009

STUDIA FENNICA ANTHROPOLOGICA

On Foreign Ground
Moving between Countries and
Categories
Edited by Minna Ruckenstein
& Marie-Louise Karttunen
Studia Fennica Anthropologica 1
2007

Beyond the Horizon
Essays on Myth, History,
Travel and Society
Edited by Clifford Sather
& Timo Kaartinen
Studia Fennica Anthropologica 2
2008

www.ingramcontent.com/pod-product-compliance
Lightning Source LLC
Chambersburg PA
CBHW080804300426
44114CB00020B/2828